Design of Composite Steel-Concrete Structures

To My Mother

Design of Composite
Steel-Concrete Structures

Lloyd C. P. Yam

Head of Structural Design Division
Building Research Station
Garston, U.K.

Surrey University Press

Published by Surrey University Press
A member of the Blackie Group
Bishopbriggs, Glasgow and
Furnival House, 14–18 High Holborn, London

British Library Cataloguing in Publication Data

Yam, Lloyd P
    Design of composite steel-concrete structures –
    (Design of structures series).
    1. Composite construction
    2. Structural design
    I. Title   II. Series
    624'.1821   TA664
    ISBN 0-903384-22-1

Printed in Great Britain by
Thomson Litho Ltd., East Kilbride, Scotland

Series Foreword

This book forms part of a series dealing with the design of structures.

The emphasis is on the word design as this is the crucial part of an engineer's activity, which distinguishes him from a scientist working in the related field. The engineer must be able to synthesize the information available for the purpose of creating on paper that which is to be built, i.e. designing. This activity requires knowledge of analysis of the whole as well as of the components of that which is being designed. It requires also knowledge of the codes and standards in force and what is considered good engineering practice. Finally, design requires engineering judgement, and judgement can come only with experience.

This is why a university course, even if design-oriented as at a few universities, does not produce a designer. The young graduate, however bright, when confronted with his first, and possibly not only first, design job simply lacks the background for his new task. This is where the books in the present series come in: they make it possible to approach the task of design in a reasonable manner.

The books cover mainly the field of structural design. Recent titles include *Glass-Reinforced Plastics in Construction* (L. Holloway) and *Design of Structural Steelwork* (P. R. Knowles). These are the sort of design problems that the civil engineer is concerned with. The books explain relatively simply the background to the design problem and then in quite some detail, the main features of the design process. All this is fully illustrated by worked examples. Such an approach may seem old-fashioned to an enthusiast of pure analysis but example and precept are essential if modern design is to build upon the accumulated stores of successful design used in the past. Indeed, modern disciplines have borrowed our approach but re-named it "case study".

The books in the present series should thus prove of great value to the young engineer and also to his slightly senior colleague who is designing in an unfamiliar field. For these people the book is a "must". But it is also a wise investment (and in these inflationary days such is not easy to come by) for the undergraduate who appreciates the importance of design. With the aid of the book he can profit much more from his university or polytechnic course and enter employment much better prepared.

The authors of the books in the series are all specialists willing, in the best spirit of the engineering profession, to share their knowledge and experience. There is therefore little doubt that the present series not only will fill a genuine need but will do so really well.

Adam Neville, General Editor
University of Dundee
May, 1980

Contents

Preface

Structural codes based on the concept of limit state design have now been published. They contain many new design recommendations but inevitably not the underlying principles. Few engineers have the time to study the relevant technical papers, even if they are readily available; furthermore, in the field of composite construction, even the research worker has to admit that the relevant information is relatively uncoordinated and incomplete. The main objective of this book is to explain the behaviour of composite structures in terms of simple material properties, thus enabling the reader to develop an ability to predict structural behaviour at the various limit states.

In preparing this book, I have assumed that the reader has no previous experience in limit state design or composite construction. Undergraduates and engineers will therefore find it suitable as a preparation for degree and professional examinations. Experienced engineers will also find the book useful because new concepts are developed from first principles and permissible-stress design is included. In many important topics, sufficient background research information is given so that the research worker can identify the needs for any further research.

This volume covers composite buildings and bridges. The principles of composite construction are briefly discussed in chapter 1. In view of the persistent co-existence of codes of different generations, a comparison is made of permissible-stress, load-factor and limit-state methods of design (chapter 2). Since structural codes and standards in the United Kingdom are increasingly influenced by their international counterparts, a state-of-the-art review on the latter is included in chapter 2. The basic behaviour of a composite structure is illustrated by simply supported T-beams in chapter 3, and equations are derived to predict their behaviour when the interaction between steel and concrete is absent, partial and complete. Continuous composite beams are discussed in chapter 4, which covers both elastic and plastic methods of design. Following a brief review of the relevant experimental and theoretical work, guidance is given on the applicability of the simple plastic theory to continuous composite beams. Chapter 5 deals with types of shear connector, including profiled steel sheeting. Local stresses around a connector are described in detail to assist the understanding of the relevant recommendation in codes.

There are miscellaneous problems relevant but not unique to composite construction. Some textbooks give general principles for their solutions, but the application to composite structures is by no means straightforward. These problems are dealt with in chapter 6, in which general equations are formulated for the analysis of the effects of shrinkage, creep and differential temperature. Methods of composite column design are developed progressively in chapter 7—from pin-ended columns with axial load to restrained columns with biaxial bending. Following a rigorous treatment, simple methods of checking column strength are provided which are sufficiently accurate for practical purposes. The final chapter contains worked examples on simply supported beams, columns under various loading conditions and a highway bridge deck of continuous construction.

I wish to express appreciation to my colleagues at the Building Research Station and the Transport and Road Research Laboratory who have assisted me directly or indirectly in the preparation of this book. Every effort has been made to eliminate errors and ambiguities from the text, but I would be most grateful to readers who draw my attention to any need for clarification or correction.

LLOYD YAM

Notation

A	area
a	spacing of connectors
α	composite stiffness factor (in beams)
	concrete contribution factor (in columns)
B	width of slab (of a T-beam)
	beam spacing
b	effective width (for flange of a beam)
	width of column
b_f	width of top flange of I-beam
	ratio of smaller end moment to larger (for columns)
	a constant
C	compressive force in slab
D	slab thickness
	outside diameter of circular column
d	depth of steel beam
d_c	distance between centroid of steel beam and midplane of slab
	half depth of composite column
δ	deflexion
E	Young's modulus
e	strain difference at interface (without suffix)
	strain (with suffix)
F	design strength of a connector
f	strength (steel or concrete)
f_y	yield stress of steel
ϕ	creep coefficient
h	overall depth of composite beam
I	moment of inertia
K	connector stiffness (force per unit slip)
k	curvature
L	span

l half span
    length of column
λ slenderness parameter
M bending moment
M' positive value of hogging moment
m ratio of maximum span moment to support moment
N number of connectors
    axial load
n depth of neutral axis
    ratio of number of connectors used to that required to develop the
    maximum moment
P magnitude of a point load
Q force on a connector
q interface shear force per unit length
R ratio of f'_{cu} to f_y
r radius of gyration
s slip
σ stress
T tensile force in steel section
t web thickness of steel section
t_f flange thickness of I-section
V vertical shear (moment gradient)
w a constant
    crack width
x distance
y distance
    depth of neutral axis (columns)
z section modulus

Suffixes
c concrete
    creep effects
cu cube strength
E Euler load
f full-connexion design
    free strain due to shrinkage, creep and temperature
    flexural strength (only in cased-strut method)

p	fully plastic value
	partial-connexion design
r	reinforcement
s	steel
	shrinkage of concrete
sh	strain-hardening
t	temperature
u	at ultimate load condition
y	at yield

1 Introduction

1.1 Introduction

In civil engineering construction, the merits of a material are based on factors such as availability, structural strength, durability and workability. It is hardly surprising that no naturally occurring material is known which possesses all these properties to the desired level. The engineer's problem therefore consists of an optimization involving different materials and methods of construction, with the objective of building a structure at minimum cost to meet his requirements.

1.2 Definitions

Methods of improving material utilization can be classified into two categories. The first is to select appropriate materials to form a new product with desired properties, thus resulting in a *composite material*; for example, glass fibres, cement and an additive have been combined to form a relatively cheap product used as cladding panels in buildings. This composite derives its tensile strength from the glass fibres and its compressive strength from the cement matrix, while its durability is appreciably improved by the use of the additive.

Alternatively, different materials can be arranged in an optimum geometric configuration, with the aim that only the desirable property of each material will be utilized by virtue of its designated position. The structure is then known as a *composite structure*, and the relevant method of building, *composite construction*. An illustrative example is the composite hybrid beam[1] shown in figure 1.1. This simply supported beam is subjected to static load causing bending and shear. The bending moment is mainly resisted by the compressive force in the concrete and the tensile force in the steel of the bottom flange. It can thus be seen that the appropriate properties of concrete and steel are effectively utilized, and that these materials are strategically placed (top and bottom) to maximize the moment arm for resistance against bending. As for the shear force, the vertical web is the main active component; hence a material with high allowable shear[2] is used. Since the top flange of the I-beam has negligible effect on the strength of the structure, only a minimum of material is provided here for the connexion of the I-beam to the concrete.

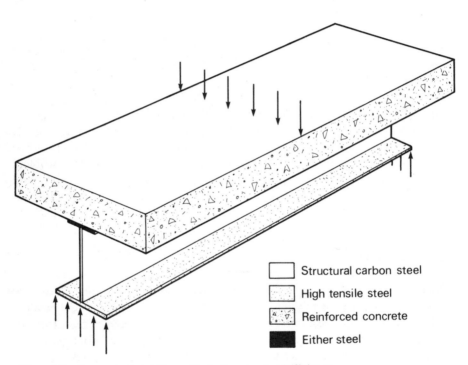

□ Structural carbon steel

▢ High tensile steel

▨ Reinforced concrete

■ Either steel

Figure 1.1   Composite hybrid beam illustrating structural efficiency

1.3 Composite action

It should be noted that the structure in the above example can be considered composite only in so far as the various components are connected to act as a single unit. The structural performance depends on the extent to which composite action can be achieved. This is illustrated in figure 1.2, which shows the reduction in deflexions and strains due to composite action. The presence of interface slip (shown as displacement of the vertical marks between beams) can also be observed from the figure. It follows that rigid connexions at the interface tend to eliminate slip and hence ensure composite action. In practice, such connexions are provided by shear connectors embedded in both components.

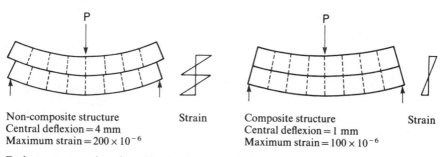

Non-composite structure   Strain   Composite structure   Strain
Central deflexion = 4 mm   Central deflexion = 1 mm
Maximum strain = 200×10^{-6} Maximum strain = 100×10^{-6}

Each structure consists of two identical beams marked at equal intervals

Figure 1.2   Composite and non-composite beams under equal loads

1.4   Scope of book

From the above definition, it can be inferred that a large number of composite structures can be produced by the combination of different structural components, including rolled steel beams, built-up sections, timber beams, precast and *in situ* concrete beams, slabs, steel plates with or without stiffeners, walls, steel tubes and frames. The scope of this book will be confined to composite steel-concrete structures, which are of particular interest to the engineer engaged in the design of buildings and bridges. In this context, the term *concrete* refers to plain or reinforced concrete, while the term *steel* includes various forms of built-up sections. Some practical examples are shown in figure 1.3.

This limitation of scope is imposed merely for the sake of convenience of presentation. The fundamental principles of analysis can nevertheless be applied to other forms of composite construction, provided that the stress-strain relations of the relevant materials are known.

1.5   Advantages and disadvantages

In general, the merit of any form of construction in civil engineering is measured by the cost of the overall structure and its subsequent maintenance. This method of assessment has become increasingly complicated in recent years, owing to a substantial increase, particularly in Europe and America,

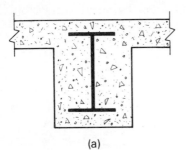

(a)

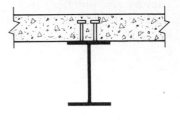

(b)

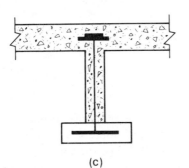

(c)

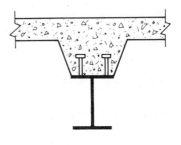

(d)

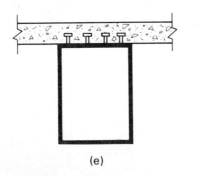

(e)

(f)

Figure 1.3 Types of composite section: (a) encased steel beam, (b) composite T-beam, (c) preflex beam, (d) composite beam with haunched slab, (e) steel box girder with composite deck, (f) composite column

in the proportion of labour cost. It is no longer valid to consider the structural cost in isolation. Due account should be taken of construction methods relevant to the particular site, and contractor experience, together with the availability of the relevant materials and labour. In view of these factors, only the qualitative aspects of composite construction are given below. When the overall situation is sufficiently defined, these items can be quantified for cost comparison purposes.

Advantages Compared with its non-composite counterpart with the same components, a composite structure tends to have greater stiffness, higher load capacity against material damage, and also higher collapse capacity. Consequently a composite section is generally smaller than alternative designs to sustain the same load, thus resulting in the saving of material, weight, and headroom or construction depth. Compared with reinforced concrete construction, composite construction allows the use of prefabricated units and has the advantage of providing facility (e.g. steel beams in bridges) for the support of the soffit shuttering.

Disadvantages The obvious difference is the need to provide shear connectors and the necessary labour involved. Compared with reinforced concrete, a composite bridge deck incurs the subsequent cost of steel maintenance, i.e. painting.

2 Concepts of Structural Design

2.1 Purpose of structural design

In broad terms, the purpose of structural design is to satisfy some functional requirements within the constraints of resources and time. Functional requirements are related to the people for whom the structure is constructed, and are therefore expressed in terms of user activities. Some major requirements[3] for a building are shown in table 2.1, but it should be noted that there may be additional requirements originating from owners, financiers, neighbours, the community or future users—concerned with safety, health, comfort and economy.

To fulfil these requirements, a structure must perform satisfactorily, i.e. remain safe and serviceable during its intended lifetime. The structural engineer has a significant role to play in ensuring such a satisfactory performance of the structure. His main concern is in designing a structure with an acceptable margin of safety against various limiting conditions of usefulness known as *limit states*. Limit states normally considered in structural design are:

1. Collapse, buckling or overturning
2. Excessive deformation (including movement)
3. Local damage (including cracking and corrosion)
4. Excessive vibration or oscillation
5. Fatigue or brittle fracture
6. Loss of durability
7. Rapid loss of strength or resistance when subjected to fire
8. Disproportionate damage as a result of gross misuse or accident

2.2 Design methods and codes of practice

Although methods of design are primarily applications of structural principles, the choice is in fact somewhat limited by the existence of statutory control systems. Four systems of building control are in operation in the

Table 2.1 Functional requirements for a building

Functional requirements	Means of satisfying requirements
Accommodation for people/goods	Floors Headroom
Access for people/goods	Entrances and openings Corridors/circulation areas and routes Ramps/stairs/escalators Lifts/paternosters/hoists
Shelter from elements	External walls Roof
Provision of light	Natural light Artificial light
View to/from outside	External openings
Supply of fresh air Removal of stale air	Natural ventilation Artificial ventilation
Control of temperature	Insulation
	Generation—within premises remote generation
Fresh water	Incoming supply pipe(s)

Removal of waste	Localized—solids/liquids
	           chutes/drains
	Dispersed—Vapours: natural draught
	                artificial extract

Privacy	Horizontal visual/acoustic separation Vertical acoustic separation
Control of external noise	External acoustic insulation
Scope for change of requirements	Flexibility for relocation of internal walls and openings through floors

United Kingdom. They cover England, Wales, Scotland, Northern Ireland and Inner London. In spite of the variation in the method of control, the bases of structural design are very similar. In England and Wales, the design of a building has to comply with the Building Regulations written under the Public Health Acts of 1936 and 1961, and the 1974 Health and Safety at Work Act. The principal regulation for structural safety is D8, which requires that a structure shall safely sustain and transmit to the foundations the combined dead load, imposed load and wind load.

For practical application, regulations have to be elaborated in engineering terms. This is done by an independent body, the British Standards Institution (BSI), which produces structural codes of practice. These codes are deemed to satisfy the regulations, but are not the only means to obtain approval. Thus, if an applicant uses such a code as the basis for Building Regulations approval, then the Local Authority cannot reject the application if it complies with the code. On the other hand, the Authority cannot insist on a particular code being used. Generally, however, codes of practice are used except in rare circumstances where special methods of calculation are needed. It should be noted that the Loading Code (CP3, Chapter 5, Part 1 and Part 2) is mandatory.

In the design of bridges, methods tend to be more standardized or even centralized, since the Government (Department of Transport) is the principal client for the nation's bridgeworks. Here again BSI codes form the basis of structural design. It thus appears that the use of structural codes constitutes a major part of the design procedure. While accepting this as inevitable, it should be stressed that mere compliance with the relevant code is insufficient to ensure a safe design. A satisfactory design is an appropriate blend of theory, practice, imagination and good judgment.

Structural codes have been subjected to continual changes during the last 50 years in response to the advance of construction technology and of the understanding of structural behaviour. This destiny of changes will continue into the 1980s, with perhaps another dimension created by the force of international harmonization. Because of the difficulty in co-ordinating the various structural codes (such as the Bridge Code and building codes for various materials), it is inevitable that their revisions are out of phase with one another. As a result, there will be generation gaps among existing codes for some time to come, and it is important for the structural engineer to understand the basic principles underlying the relevant codes. Accordingly, the development of these underlying principles is discussed in the following sections. A list of the major codes relevant to buildings and bridges is given in table 2.2.

2.3 Permissible stress design

The loading used in this method of design is known as the *working load* and is intended to represent the highest load which may be expected to occur under normal working conditions. Values of working loads are obtained from the relevant codes or, in the case of dead loads, calculated from volumes and densities. To calculate the effects of loads on a member or structure, these loads (dead, imposed, wind and temperature loads) are combined, without the use of any multipliers (i.e. unfactored), to obtain the most adverse design values of moments and shears in a particular member or at a particular point.

Table 2.2 Major structural codes

CP3: Chapter V, Part 1	Dead and imposed loads
CP3: Chapter V, Part 2	Wind loads
CP 114	The structural use of reinforced concrete in buildings
CP 115	The structural use of prestressed concrete in buildings
CP 116	The structural use of precast concrete
CP 110	The structural use of concrete
BS 449	Use of structural steel in building
BS 153	Steel girder bridges
CP 117	Composite construction in structural steel and concrete
	Part 1: buildings
	Part 2: bridges
CP 112	The structural use of timber
BS 5628	Structural use of masonry
	Part 1: unreinforced masonry
CP 2003	Earthworks
CP 2004	Foundations
BS 5400	Steel, concrete and composite bridges
DD 55	Fixed offshore structures
BS 5502	Design of buildings and structures for agriculture
BS 2053	General-purpose farm buildings of framed construction

Stress is used as the basic parameter to indicate the margin of safety, and as a rule elastic analysis is used to locate and evaluate the maximum stress. Maximum stresses must not exceed certain limits (specified in codes and known as *permissible stresses*) which, in the case of concrete, are about 50% of the flexural strength and, in the case of steel, are about 60% of the yield strength. The criterion can be expressed in terms of the *safety factor* (SF) as follows:

$$\text{working stress} \leqslant \text{permissible stress}$$
$$\text{permissible stress} = \text{ultimate (or limiting) stress} \div \text{SF}$$

This method is admirably simple and has been a satisfactory basis of structural design for a long time. It seems obvious that the validity of using stress level at a point to characterize safety is bound to be limited. It is therefore natural to expect the permissible-stress approach to be less meaningful in dealing with plastic collapse and instability. A separate ultimate load method is necessary and, in the design of steel framework, desirable because the calculation is simpler and the design is more economical. The design method based on the evaluation of ultimate loads is known as the *load factor method* and was incorporated into structural codes in the late 1950s.

2.4   Ultimate load design (load factor method)

There are variations of this method depending on the structural material and the stage of development of the relevant code. The basic objective is to provide an additional parameter to assess the margin of safety. This is the ratio of the ultimate load of the structure to the working load and is referred to as the *load factor* (ultimate load = load which would cause collapse).

The working loads are the same as those used in the permissible-stress design. They are then combined to form a group of loads known collectively as the *combined working load*, which is critical to a given structure associated with a given collapse mode. If the value of the combined working load is increased until the structure collapses, the level of load attained is called the *ultimate load*. The magnification factor (i.e. ratio of ultimate load to working load) is known as the *load factor* (LF) and is about 2.0. The material strengths used for the collapse analysis are the ultimate strengths. This ultimate load criterion can be expressed as follows, where the load factor is specified in the relevant code:

$$\text{working load} \times \text{LF} \leqslant \text{ultimate load}$$

2.5   Limit-state design

2.5.1   *Limit states*
A study of the development of concrete codes since 1934 has shown that safety factors and load factors have decreased steadily.[4] The advancement in concrete technology has brought about a wider range of its application and enabled lighter structures to be built. Similar developments in steel construction can also be observed; the post-war advancement in welding techniques has resulted in the use of thin-plate structures, culminating in the construction of long-span bridges of relatively low weight. The structural engineer of today has to be more specific in checking the likely performance of his designed structure. The limiting conditions of overstressing and collapse discussed previously are only two of many possible limit states with which he may be concerned (see 2.1). Even though the number of limit states in a particular case is often small, a rational approach is needed to deal with the various limit states on a common basis.

2.5.2   *Partial safety factors*
One possible solution is to adopt the load factor method. Since the required margin of safety against collapse is very different from that against local damage, the use of different factors for different limit states appears to be reasonable. These load factors, however, do not take account of the variability

of material properties. Furthermore, any such material factor should be applied to material strength rather than loading, so that a value appropriate to a specific material can be used.

The framework of limit-state design provides a satisfactory solution to the above problems. Firstly, the global safety factor is split into two parts— partial safety factors for loads (as load multipliers) and partial safety factors for materials (as strength divisors). Secondly, there is generally one set of partial safety factors for the ultimate limit state (i.e. collapse) and another set for the serviceability limit states (other than collapse). Since the design strength is obtained by dividing a certain specified strength by the appropriate partial safety factor, this factor has no unique significance until the specified strength is defined. In British codes, this strength is defined in statistical terms as the *characteristic strength*, i.e. that strength below which only a given percentage of the test results will fall. A 95% fractile has been adopted by CP110, implying that not more than 5% will fall below the characteristic strength. Similarly, characteristic loads are used in conjunction with the relevant partial safety factors to obtain the design loads.

2.5.3. *Reliability theory*

So far, what has been established is a rational and convenient framework which is more suited to deal with the wide range of uncertainties associated with the assessment of various limit states. The next question is naturally how to evaluate these partial safety factors to achieve the desired margins of safety. Strictly, this question is mainly the concern of code committees but, in order to enhance the understanding of the significance of the limit-state calculations, a qualitative discussion of the basic principles is given here. It will be convenient to describe the *reliability theory* first.

If the collapse load of a structure is a function of several parameters, such as material strength and areas of cross-sections, then the collapse load has a deterministic value when all the relevant parameters are known with certainty. If the parameters are not represented by single values but by frequency distributions (e.g. results of cube tests), then the collapse load will also be represented by a frequency distribution. When the frequency distribution of the loading is also known, the application of reliability theory will enable the probability of collapse to be evaluated. Thus, when the parameters are deterministic, the margin of safety can only be expressed as a resistance/load ratio; when the frequency distributions are available, the margin of safety can be expressed as risk of failure.

Assuming that all the relevant uncertainties could be modelled mathematically, and that the required data were available, application of the reliability theory can enable the appropriate partial safety factors to be evaluated so that, for a given limit state, the risk of occurrence has a prescribed value when the various variabilities are taken into account. This meets the limit-state requirement that the risk of reaching a limit state should be acceptably remote.

From a consideration of probability, the previous methods of load combination have to be modified for the limit-state application. Because of the reduced probability that two (or more) loads will occur together at their characteristic values, a reduction in the relevant partial safety factor is required. This accounts for the different sets of partial safety factors contained in limit-state codes (three combinations in CP110 and five combinations in the Bridge Code for the ultimate limit state).

2.5.4 *Future of structural codes*

It appears that the permissible stress and load factor methods will continue to be used for some time. There is no doubt that they can give satisfactory service to the practitioner when used correctly and with good judgment. However, all major structural codes currently under review in the United Kingdom are based on the limit-state concept, and there are simply no resources available for the revision of codes on any other basis.

It is generally accepted that the limit-state approach provides a rational basis for codification and design. However, further efforts have to be made to study in detail uncertainties of practical significance and to collect the relevant data to improve the methods and figures contained in current limit-state codes. In view of the limitation of the relevant knowledge, a simplification of the design procedure recommended in current limit-state codes is justifiable. In particular, a re-examination of the load combination concept is highly desirable in order to reduce the design calculation substantially.

British codes and standards are increasingly influenced by their continental and international counterparts. International trade and overseas construction have played a part, but oddly enough not the major part, in bringing about this phenomenon. The main driving forces come from engineers convinced of the benefit of working together and government commitments towards harmonization, although these forces may be of a different nature. The young engineer's career is bound to be affected by the development of international activities related to structural design.

2.6   Development of EEC and international codes and standards

2.6.1   *Introduction*

In view of some recent efforts to accelerate the convergence of international codes and standards, this section seeks to give a brief account of the state of the art and outlines the basic framework of structural codes recommended for international harmonization. To the engineer who looks ahead, the treatment of this framework should prove particularly useful, since it is likely to become a part of the future British scene in one way or another.

Although technical harmonization is as old as engineering science, the legal consequences of harmonization have only recently begun to be explored. To limit the scope of the present discussion, only two bodies, the European Economic Community (EEC) and the International Standards Organization (ISO), need be considered as significantly influential. ISO was set up over 30 years ago under the United Nations banner. It includes more than 60 member countries and deals with a wide range of standards from ball bearings to structural codes. As a member of ISO, it is natural for the UK (British Standards Institution) to adopt ISO standards as the basis for national standards, with national variations or refinement as necessary. More recently complete adoption of international standards by BSI has become the recognized idea. This appears to be justified where the adoption of such standards would assist our exporters or amount to savings in resources. The legal status of EEC documents is more binding by virtue of the Treaty of Rome. Community Directives, once agreed, are binding on Member States and, even when applied in a milder form, would require the UK to give the relevant documents the same status as British Standards. The EEC Commission is in the process of drafting seven structural codes known as *Eurocodes*. The first Eurocode deals with design principles for the various material codes, and can thus be regarded as a master code for the remaining six Eurocodes which deal with concrete, steel, composite construction, masonry, timber and foundations. All Eurocodes will be based on the limit-state philosophy and will cover buildings and bridges. Before implementation, adequate time is expected to be required for consultations within Member States and voting by the Council of Ministers. Table 2.3 compares the *status quo* in respect of structural codes in six EEC Member States. The author is unable to obtain the relevant information on Belgium, Luxembourg and the Republic of Ireland.

2.6.2 *Common basis for international structural codes*
All international model codes in existence are based on the limit-state concept and, because of the early existence of a master code coupled with a strong link among the various international bodies, these model codes are as well co-ordinated as could be expected of any multi-national groups. The partial safety-factor formats are substantially consistent. As far as structural calculation is concerned, these codes do not differ in principle from the British counterparts (current or under preparation). This is particularly true in the case of the model code on composite structures, which has been given strong British support, in terms of both participation in drafting and research

Table 2.3 Codes of practice—simplified table of differences among some EEC countries

	UK	Germany	Denmark	France	Netherlands	Italy
(A) Who issue Regulations & Codes	Secretary of State Department of Finance (Northern Ireland)	DIN	Ministry of Housing	Public Contracts—Ministry of Equipment French Standards Association (AFNOR)		Ministry of Public Works
	BSI	Ministers of Federal Laender Additional rules by Ministry of Transport	DIF (Danish Society of Engineers)	Private Contracts—Standards Technical Documents Group or above	NNI (Netherlands Standards Institute)	CNR, UNI (Italian Standards Institute)
(B) Are they mandatory?	(1), (2) and (3) mandatory (1) Building Regulations (2) Loading Codes (3) Standards & Technical Memoranda for Highway bridges (4) British Standards—deemed to satisfy	Yes (*de facto*) Approval may be given for departure in special cases	Laws and Circulars—Yes Codes—deemed to satisfy	Public Contracts—Yes Private Contracts—Builders subject to 10-year guarantee by law. Hence Insurance companies become approval bodies	Yes if client and contractor agree	Laws, Decrees, Circulars—Yes

	Building Regulations do not specify roles	Yes	Yes	Not generally	No	Yes, two "Directors": Resident Engineer (design/construction) Site Agent (Workmanship)
(C) Do they define roles of parties to building works?	Building Regulations do not specify roles	Yes	Yes	Not generally	No	Yes, two "Directors": Resident Engineer (design/construction) Site Agent (Workmanship)
(D) Is there a Master Code (Principles for various Material Codes)?	No	Yes, since 1977. Revised version near completion (not mandatory)	Yes, 1978 (NKB document)	Yes, since 1971	Yes, since 1972	Yes (1979)
(E) Future Codes based on Limit States?	BSI Committees agreed	Yes	Yes, since 1965	Yes, gradually	Yes, gradually	Yes, gradually
(F) Enthusiasms in Reliability Theory (0 = very low)	0	2	3	1	1	2

input. Its basic principles of structural design are therefore very similar to those of the Bridge Code (BS 5400) and hence to the contents of this book. The European concept of loads and their combination, however, is different from that adopted by British codes.

There are nine definitions of loads (also referred to as *actions*) but this classification has relatively little consequence because no new concepts are involved. Thus, loads can be described as direct, indirect, fixed, free, static, dynamic, permanent, variable and accidental. The only note which needs to be added is that permanent and variable loads are equivalent to the British dead and live loads respectively.

In respect of load values, two concepts appear to be new to the British engineer. Firstly, in addition to the characteristic load value specified in a loading code, two further values are given (in terms of two factors). This applies only to a live load when the variation of its magnitude over a given period of time is known statistically. If only the characteristic value is used in design, some higher values would be ignored which may occur frequently enough to make the structure unfit. Hence a higher value known as the *frequent value*, with a lower probability of occurrence than the characteristic value, is provided for use in design where appropriate. There is yet another value worth obtaining from the known statistics: there may be a substantial portion of the live load which remains on the structure as the load fluctuates with time. This portion of the live load acts virtually as dead load and should be considered in the calculation of long-term effects (e.g. creep). Hence another load value is taken from the known statistics of the live load for the assessment of long-term effects; this is known as the *quasi-permanent value*. The other novelty is related to load combination. Again, in addition to the specified characteristic value, a combination value is to be given in a code for load-combination calculations.

There is no denying that this type of loading code seems unnecessarily complicated from the British point of view. But, on the other hand, the British format would be inadequate to cover all types of load, or the same type of load with different natures of variability. For example, while earthquake loading is not considered in the United Kingdom, it is an accidental load in one part of Italy, and becomes a live load in another part of the same country. Snow loading in some Nordic regions could cause structural failure and is treated as the most critical load, while temperature loading in some parts of the Soviet Union is the major consideration. It is also known that wind speeds in Siberia can be so constant throughout the year as to be treated as dead load.

3 Composite Beams

3.1 Simply supported beams

3.1.1 *Full and partial interaction*

In order that the steel beam and the slab act as a composite structure, the connectors must have adequate strength and stiffness. If there are no horizontal or vertical separations at the interface (i.e. no slip or uplift), the connectors are described as *rigid*; complete interaction can be said to exist under these idealized circumstances. However, all connectors are flexible to some extent, and therefore partial interaction always exists. For most connectors used in practice, failure by vertical separation is unlikely and any uplift would have only negligible effect on the behaviour of the composite structure. It is therefore sufficient to consider only slip in the study of the effects of partial interaction.

The quantitative effects of composite action can be appreciated from a discussion of the methods of analysing stresses and deflexions. Accordingly, the following three cases of a T-beam (slab and I-beam) are used to illustrate the interaction phenomenon. Concrete is assumed to be uncracked in all three cases. The corresponding strain distributions at midspan (for the same bending moment) are also shown (figure 3.1).

Case 1   No interaction

Since there is no horizontal force at the slab-beam interface, neither is subjected to any axial force. Assuming that the concrete behaves elastically in compression and tension, the strain distributions are as shown in figure 3.1. Without vertical separation between the slab and the beam, their curvatures at any given cross-section are equal. Since the condition is equivalent to the pure bending of two beams with equal curvatures along the span, the moment-curvature relation of the composite beam is one of simple summation:

$$M = k(E_c I_c + E_s I_s) \qquad (3.1)$$

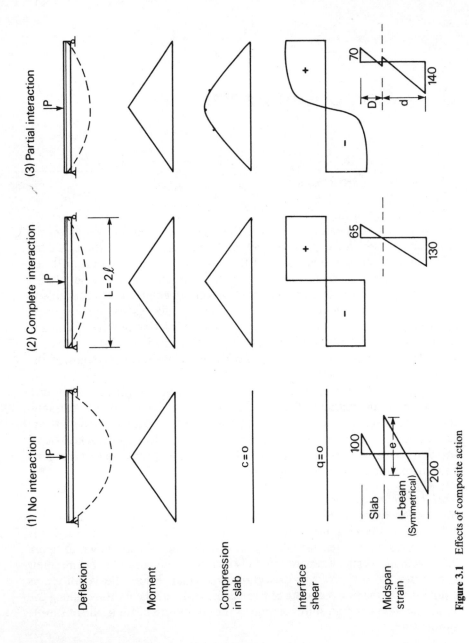

Figure 3.1   Effects of composite action

The strain difference at the interface is given by

$$e = k\left(\frac{D}{2}+\frac{d}{2}\right) \doteq \frac{M(D+d)}{2(E_cI_c+E_sI_s)} \tag{3.2}$$

The curvature and strain are uniquely defined by the external moment and the solution is very simple. To provide an early appreciation of the interaction phenomenon, it is appropriate to introduce here the calculation of slip along the span.

Such an occurrence of horizontal displacement at the interface can be explained with the aid of figure 3.2. The effect of loading is to induce tensile

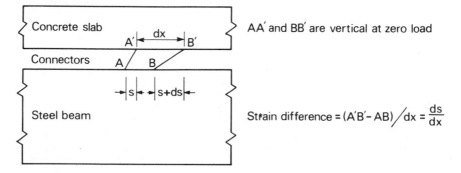

Figure 3.2   Strain difference at interface due to slip

strain at the slab soffit and compressive strain at the top surface of the steel beam. Because of symmetry, there is no slip at midspan. Moving away from midspan slip increases as a result of the strain difference e at the interface. To calculate the slip the following geometric relation derived in figure 3.2 is made use of:

$$e = \frac{ds}{dx} \tag{3.3}$$

where s (the slip) is defined as the displacement in the x-direction of a point at the slab soffit relative to the top flange of the I-beam.

The end slip can be obtained by integration, starting at midspan where slip is zero:

$$s_{\text{end}} = \int_0^{L/2} e\, dx = \frac{D+d}{4(E_cI_c+E_sI_s)} \times (\text{area of BM diagram}) \tag{3.4}$$

For the case shown in figure 3.1 (1):

$$s_{end} = \frac{(D+d)PL^2}{32(E_cI_c + E_sI_s)}$$

(3.5)

This simple formula gives the end slip for the condition of no interaction and thus provides a convenient estimate for the upper bound value of slip for the case of partial interaction.

Case 2   Complete interaction

When the interface connexion is sufficiently rigid, the strain difference due to slip may be neglected in the calculation of stresses and deflexions. Accordingly the strain distribution over the depth of the composite section consists of a continuous straight line (figure 3.3). The normal elastic calculation can be used, assuming that concrete has the same elastic modulus (E_c) in compression and tension (uncracked).

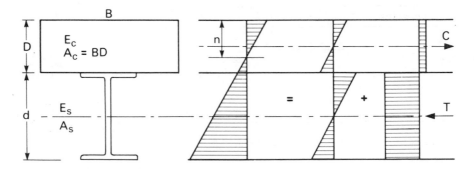

Figure 3.3   Strain distribution for complete interaction

The compressive force in the slab is

$$C = kE_cA_c(n - D/2)$$

(3.6)

This should be equal to the tensile force in the I-beam:

$$T = kE_sA_s(D + d/2 - n)$$

(3.7)

The depth of the neutral axis is obtained by equating C and T:

$$n = \frac{D}{2} + \left(\frac{D+d}{2}\right)\left(\frac{E_sA_s}{E_sA_s + E_cA_c}\right)$$

(3.8)

The moment of resistance can be calculated by resolving the forces as shown in figure 3.3, into the no-interaction part (figure 3.1 and equation 3.1) and the equal forces at the neutral axes.

$$M = k(E_cI_c + E_sI_s) + C\left(\frac{D+d}{2}\right) \tag{3.9}$$

To determine the stresses due to a given bending moment the following moment-curvature relation, derived from (3.6), (3.8) and (3.9), will be useful:

$$\frac{M}{k} = (E_cI_c + E_sI_s) + \left(\frac{D+d}{2}\right)^2 \left(\frac{E_cA_c \times E_sA_s}{E_cA_c + E_sA_s}\right) \tag{3.10}$$

The equation is rewritten in a different form below to incorporate the "composite stiffness factor" α, which depends only on the section properties and shows the extent of composite action (equation 3.1 can be obtained by setting $\alpha = 0$):

$$M = k(E_cI_c + E_sI_s)(1 + \alpha) \tag{3.11}$$

$$\alpha = \frac{E_cA_c \times E_sA_s}{E_cA_c + E_sA_s} \times \frac{(D+d)^2}{4(E_cI_c + E_sI_s)} \tag{3.12}$$

Before proceeding to the discussion on partial interaction, it will be instructive to calculate the interface shear for the present example of rigid connexion. It should be noted that the shear reduces with increasing connector flexibility, falling to zero in the case of no interaction discussed previously.

The interface shear force per unit length (q) is defined in figure 3.4, the sign convention being consistent with that for slip (figure 3.2) so that positive

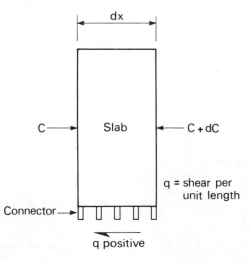

Figure 3.4 Equilibrium of slab element

shear corresponds to positive slip. From figure 3.4, consideration of equilibrium gives

$$q\,dx + dC = 0 \tag{3.13}$$

Hence

$$q = -\frac{dC}{dx} \tag{3.14}$$

To calculate the distribution of q along the span, an expression for C is derived and then differentiated. For the present example, the required expression is obtained by eliminating k and n from equations 3.6, 3.8 and 3.9:

$$M = \frac{(D+d)(1+\alpha)}{2\alpha}\,C \tag{3.15}$$

Hence the interface shear is constant along the span and is given by

$$q = \frac{2\alpha}{(1+\alpha)(D+d)}\left(-\frac{dM}{dx}\right) = \frac{2\alpha V}{(1+\alpha)(D+d)} \tag{3.16}$$

This is plotted on figure 3.1. It can be seen that equation 3.16 provides a convenient means for an estimate of the upper bound of shear when incomplete interaction occurs.

Case 3   Partial interaction

When the occurrence of slip induces appreciable strain difference at the interface, the equilibrium analysis of the section has to take account of the strain difference. As a result, the strain distribution can no longer be determined from a given bending moment, as in the previous two cases (equations 3.1 and 3.10). The following derivation will show that the

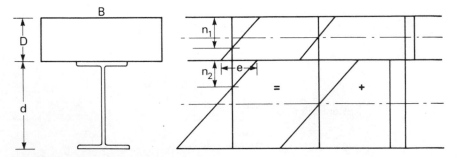

Figure 3.5   Strain distribution for partial interaction

compressive force in the slab depends on both the moment and the strain difference (equation 3.21).

From figure 3.5, the forces, the bending moment and the strain difference can be written as:

$$C = kE_cA_c(n_1 - D/2) \qquad (3.17)$$

$$T = kE_sA_s(d/2 - n_2) \qquad (3.18)$$

$$M = k(E_cI_c + E_sI_s) + C(D/2 + d/2) \qquad (3.19)$$

$$e = k(D + n_2 - n_1) \qquad (3.20)$$

Equating C and T and eliminating k, n_1 and n_2, an equation comparable to (3.15) is obtained:

$$M = \frac{(D+d)(1+\alpha)}{2\alpha}C + \frac{2(E_cI_c + E_sI_s)}{D+d}e \qquad (3.21)$$

In order to reduce the two unknowns (C and e) to one, the strain difference in equation 3.21 will be expressed in terms of C using equations 3.3 and 3.14. In order to do so, the missing link between the shear q and the slip s must be provided. This is done by the use of the elastic stiffness of a connector K, defined as the shear force on the connector to produce unit slip. It follows that

$$K = qa/s \qquad (3.22)$$

where a = spacing of connectors. Hence

$$e = \frac{ds}{dx} = \frac{a}{K}\frac{dq}{dx} = \frac{a}{K}\left(-\frac{d^2C}{dx^2}\right) \qquad (3.23)$$

Substituting the above equation into (3.21), the following governing differential equation is obtained for the analysis of the partial interaction of an elastic composite beam with uniform connector spacing.

$$\frac{2a(E_cI_c + E_sI_s)}{K(D+d)} \cdot \frac{d^2C}{dx^2} - \frac{(D+d)(1+\alpha)}{2\alpha} \cdot C + M = 0 \qquad (3.24)$$

The two boundary conditions are that C vanishes at both end supports. To solve the above equation for the loading shown in figure 3.1, the left support can be taken as the origin and, because of symmetry, only half the span is considered. The bending moment M can then be replaced by $Px/2$. To simplify the writing of equations, the following form will be used for equation (3.24), where w and β are constants calculated from the given structural data.

$$\left.\begin{aligned} \frac{d^2C}{dx^2} - w^2C + w^2\beta x &= 0 \\[2mm] w^2 &= \frac{(D+d)^2}{4(E_cI_c + E_sI_s)} \cdot \frac{K(1+\alpha)}{a\alpha} \\[2mm] \beta &= \frac{P\alpha}{(D+d)(1+\alpha)} \end{aligned}\right\} \tag{3.25}$$

A solution satisfying $C=0$ at the support and $dC/dx=0$ at midspan is

$$C = \beta\left(x - \frac{\sinh wx}{w \cosh wl}\right) \tag{3.26}$$

From this solution, the distributions of curvature (k) and of neutral axis depth (n_1) can be calculated, using equations (3.19) and (3.17) respectively. Some results are plotted in figure 3.1.

3.1.2 *Ultimate load as basis of design*
In the above elastic analysis, some of the formulae appear to be sufficiently simple to be used in design. But any analysis involving partial interaction tends to be complicated, and it seems likely that a shear connexion design based on such analysis would be equally intractable. In practice, the composite section and the shear connexion (particularly the latter) are designed to satisfy the ultimate limit state, and the serviceability limit states are subsequently checked by simple elastic calculations.

With complete interaction, a typical stress distribution at ultimate (maximum) moment is shown in figure 3.6(E). To simplify calculations, the use of idealized stress blocks is satisfactory, provided that a reduced cube strength is taken as the average concrete stress. The strength of any reinforcement can be taken into account, but such a contribution is usually negligible. From figure 3.6(A) the ultimate moment (M_p) can be derived as follows:

Case A
$$C = Bnf'_{cu} = A_sf_y \tag{3.27A}$$

$$M_p = A_sf_y(d/2 + D - n/2) \tag{3.28A}$$

If the plastic neutral axis lies below the slab, the stress blocks should be as shown in figure 3.6(B) or (C). The neutral axis is again determined by equating the compressive and tensile forces, and the moment (M_p) conveniently calculated from a modified diagram such as figure 3.6(D), obtained by adding a pair of equal and opposite forces.

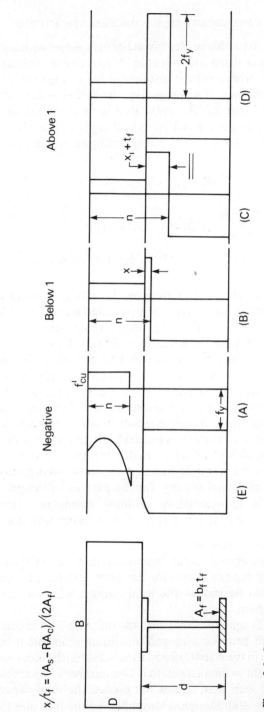

Figure 3.6 Stress distributions at ultimate load

It would be useful to be able to identify at the very outset which of the three cases of neutral axis position is applicable. A convenient indicator-formula is given in figure 3.6, where x is the distance of the neutral axis (NA) below the steel-concrete interface. If x is negative, the NA is in the slab (Case A) and equations 3.27A and 3.28A apply. If x is positive but less than the thickness of the steel flange, the NA is in the flange (Case B). If x exceeds t_f, the NA is in the web (Case C). The corresponding equations are:

Case B
$$n = D + x \tag{3.27B}$$
$$M_p = f_y(A_s d_c - n x b_f) \tag{3.28B}$$

Case C
$$n = D + t_f + x_1 \tag{3.27C}$$
$$M_p = f_y[A_s d_c - A_f(D + t_f) - t x_1(n + t_f)] \tag{3.28C}$$
where
$x_1 = (A_s - R A_c - 2A_f)/(2t) = $ depth of plastic web
$R = f_{cu}/f_y$

The effect of incomplete interaction on the ultimate moment may be considered as a reduction in capacity due to an interface strain difference. An analysis of this effect is a complicated problem, since concrete, steel and the shear connectors will behave inelastically at ultimate load.

For the purpose of design, it is sufficient to note that the reduction in ultimate moment due to the occurrence of slip can be neglected. This can be appreciated from figure 3.7, which shows the ultimate loads of a series of composite beams with varying connector strength.[5] It can be observed that, as the number of connectors in the composite beam is reduced, the ultimate load decreases slowly until point A is reached, when a more noticeable drop can be observed. It should be noted in this instance that it is the shear connexion that fails, the ultimate moment capacity having only partially developed. It is therefore satisfactory, for the purpose of design, to assume complete interaction in calculating ultimate moments, provided that adequate connexion is used to prevent premature connector failure.

3.1.3   *Design of shear connexion*
The various types of connectors and their characteristics will be discussed in chapter 5. When the type of connector has been selected, design will then consist mainly of two decisions—the total number of connectors and the manner of spacing them.

A simple method is again possible with the ultimate-load approach. First, a length of the beam between zero and maximum moments is considered. The maximum force in the slab (C in equation 3.27) for the cross-section with the maximum moment is then calculated. The number of connectors within this length must be such that C does not exceed the total capacity of the connectors (see figure 3.8). Design is simplified by the fact that the spacing

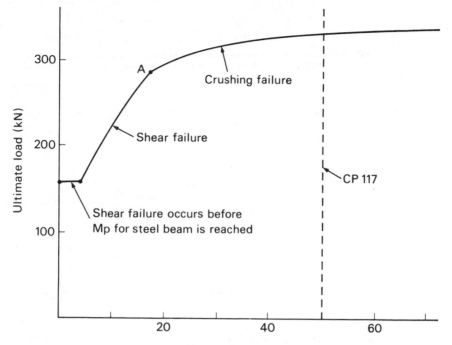

Figure 3.7 Effects of connexion strength on ultimate load (point load at midspan). Redrawn from Yam, L. C. P. and Chapman, J. C., "The inelastic behaviour of simply supported composite beams of steel and concrete", *Proceedings of the Institution of Civil Engineers*, Vol 41, Paper 7111, December 1968, pp. 651–683

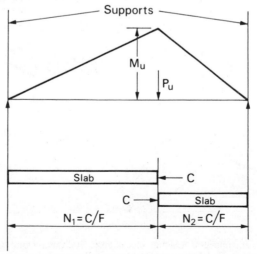

Figure 3.8 Method of shear connexion design

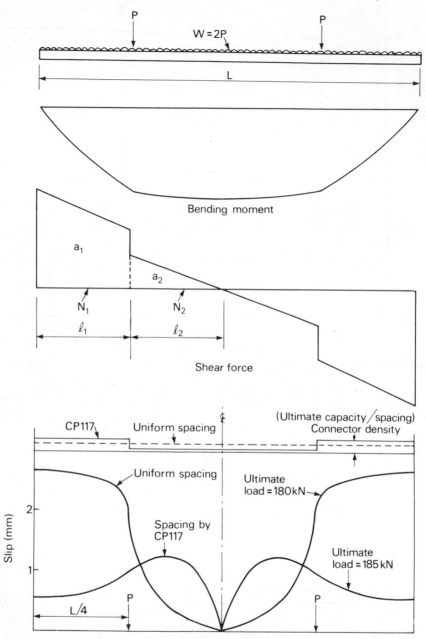

Figure 3.9 Effects of connector spacing on collapse. Redrawn from Yam, L. C. P. and Chapman, J. C., "The inelastic behaviour of simply supported composite beams of steel and concrete", *Proceedings of the Institution of Civil Engineers*, Vol. 41, Paper 7111, December 1968, pp. 651–683

has little effect on the static ultimate load of the composite beam. A uniform spacing has therefore become the standard practice. It is interesting to note that, in spite of a non-uniform elastic shear distribution, a uniform connector spacing can be used without significant reduction in the ultimate load (figure 3.9). It should be further noted that the ultimate load is insensitive also to connector flexibility, thus making it unnecessary to consider the latter in design, as is illustrated in figure 3.10.

3.2 Effective width of beam-slab system

The above analysis has so far been confined to isolated composite T-beams. A typical form of composite construction, such as a floor within a building or a bridge deck, consists of a slab connected to a series of parallel beams. Although it is valid in principle to divide the system into a series of T-beams, the slab width obtained from such a simple division may not be fully effective in resisting compressive forces. Figure 3.11 compares the deformations of two identical slabs subjected to uniform compression and point forces respectively. It can be seen that the concrete near the edges in case (b) is less effective (has lower stresses). Higher shear is induced near the horizontal centre line, but falls off in the direction y. This phenomenon of reducing shear with distance is known as shear lag. In the beam-slab shown in figure 3.12, the transmission of shear from the connectors on the top flange of the steel beam to the slab becomes less effective as the beam spacing (**B**) increases. Consequently the longitudinal compressive stresses at the top of the slab have a non-uniform distribution, as shown in the figure. In order that the T-beam approach can be used, a reduced value of the width of slab, termed effective width (b), is therefore used in design.

In addition to the geometrical dimensions of the composite system, the effective width is affected by various other factors, such as the type of loading and the support conditions. The effects of two major factors of importance, the ratio of beam spacing to span and the type of loading, are shown in figure 3.13.

In most Codes of Practice, very simple formulae are adopted for the calculation of effective widths. The formulae used by various codes are shown in table 3.1 and a numerical comparison made in figure 3.14. More extensive data are given in the Bridge Code (BS 5400: Part 5: 1979), with different values for the support, quarter-span and midspan sections respectively. In addition to tables containing effective width ratios for simply supported, cantilever and continuous beams, a general method is also given in the Appendix of BS 5400.

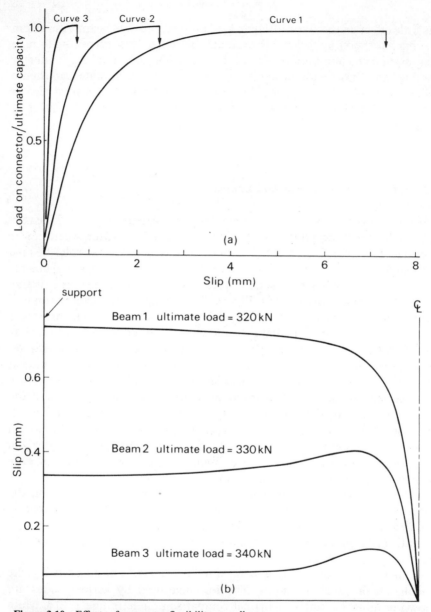

Figure 3.10 Effects of connector flexibility on collapse

    (a) Load-slip curves for three types of connector

    (b) Slip distributions at collapse (beams with equal connexion strength but different types of connector). Redrawn from Yam, L. C. P. and Chapman, J. C., "The inelastic behaviour of simply supported composite beams of steel and concrete", *Proceedings of the Institution of Civil Engineers*, Vol. 41, Paper 7111, December 1968, pp. 651–683

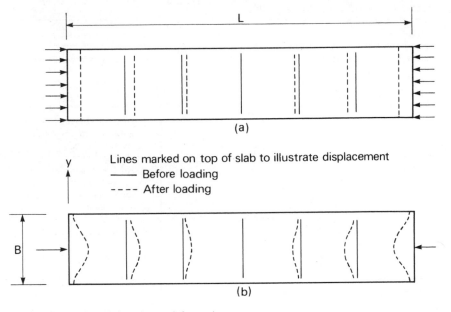

Figure 3.11   Effect of shear lag on deformation

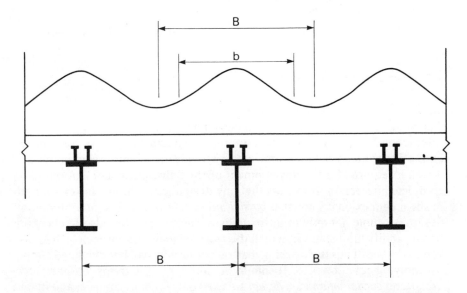

Figure 3.12   Distribution of longitudinal stresses in slab

Table 3.1 Formulae for effective widths

Code	Formulae	Comments
AASHO	b is least of (1) $L/4$ (2) B (3) $12D$	Code also covers edge beams
CEB	$b = L/8$ for u.d.l. $b = L/10$ for point load	$L=$ distance between points of contraflexure for continuous beams
CP 110	b is least of (1) $b_f + L/5$ (2) B	$L=$ distance between points of contraflexure for continuous beams which may be taken as 0.7 span
CP 117 Part 1	b is least of (1) $L/3$ (2) B (3) $b_f + 12D$	Code also covers edge beams
CP 117 Part 2	(1) For $B \leqslant L/10$; $b = B/2$ (2) For $B > L/10$, $$\left(\frac{B}{b}\right)^2 = 1 + 12\left(\frac{B}{L}\right)^2$$	
DIN 1078	(1) $B/L < 0.1$; $b/L = 1$ (2) $B/L = 0.1$ to 0.6; $b/L = 0.89$ to 0.5 (3) $B/L > 0.6$; $b/L = 0.3$	

3.3 Partial-connexion design

3.3.1 *A rational approach to design*

The method of the shear connexion design outlined in 3.1.3 is known as full-interaction design—so called because the maximum shear force transmitted by the connectors would be sufficient to provide that compression in the slab which is required for the development of the fully plastic moment based on complete interaction. It follows that any design resulting in a lower strength of shear connexion is known as *partial-interaction design*. The full-interaction design is simple not only in ultimate load calculations, but also in checking serviceability limit states in which the section modulus approach (3.1.1, case 2) can be used. Thus the working load on connectors can be calculated simply by applying equation 3.16. It should be noted that the terms *full-interaction design* and *partial-interaction design* are used only for convenience, and in fact may be misleading if taken literally. There is no ambiguity when applied to

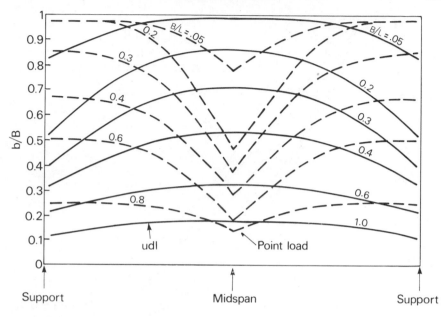

Figure 3.13   Variation of effective width (theoretical results)

structural behaviour, because partial-interaction and full-interaction simply imply "with slip" and "without slip" respectively. With shear connexion, however, no sensible connexion design can be identified which defines such a demarcation. In fact, the only logical demarcation in respect of shear connexion design is "critical shear connexion", which draws a line between beams associated with connexion failure and those associated with flexural failure. All designs beyond this critical value will have flexural failures only, with slightly higher ultimate moments as the number of connectors increases (see figure 3.7). In this book, the terms *full-connexion design* and *partial-connexion design* will be used in place of full-interaction design and partial-interaction design.

The full-connexion design was introduced into CP 117 in 1965 and has since become the preferred method in the United Kingdom and abroad for shear connexion design. For a long time, it has been recognized that this method tends to be over-conservative in a majority of cases where fatigue is not a problem. The recent introduction of the partial-connexion design into Codes of Practice, although limited at present to simple beams in buildings, is nevertheless a significant advance towards rational and economical design. This came about mainly because of work by Johnson,[6] who devised various simplified methods of design. In addition to cost saving, partial-connexion design alleviates the problems of connector congestion, enabling the use of

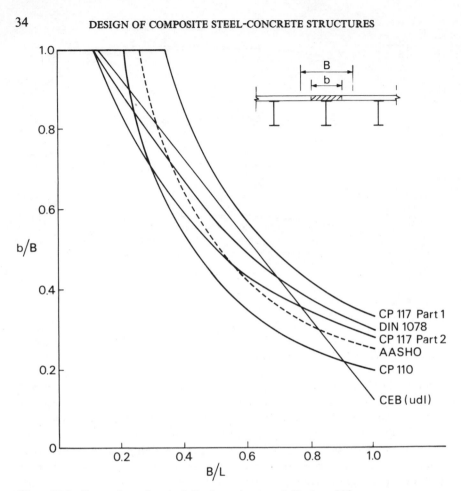

Figure 3.14   Comparison of methods for determination of effective width

a wider connector spacing and a corresponding reduction in the amount of transverse reinforcement in the slab. This is often a great help in detailing, especially when tying of precast floor slabs is involved.[7] When the slab is cast on metal decking with corrugations perpendicular to the flange of the steel beam, the voids under the corrugations would reduce the effectiveness of shear transmission from the connectors. In this case, the use of partial-connexion design is almost inevitable.

As far as efficiency in achieving ultimate capacity is concerned, full-connexion design seems both unnecessary and inefficient in many practical designs. It is inefficient because a 100% increase of shear connexion may result in only a few percent of increase in the ultimate flexural capacity. It is in many instances unnecessary, because the design of the concrete slab and of the steel beam may be governed by serviceability considerations or other criteria. In the design of multi-storey multi-bay buildings, an optimum solution may result in a large number of steel beams of different sizes. In practice, only limited sizes would be used, even if some members are much more than adequate, for architectural reasons or for the ease of construction. In these cases, the over-size members would not be called upon to develop the fully plastic moments, and it is more relevant to ask what shear connexions will be required to develop the reduced ultimate strength to resist the applied moments.

A rational approach to the discussion of an optimum shear connexion is to observe the performance of a series of composite beams with varying numbers of connectors. As far as short-term static loads are concerned, performance is characterized by collapse and serviceability behaviours, the latter covering deflexion, loads on connectors, and concrete and steel stresses at working loads. From figure 3.7, which shows the types of failure and the corresponding ultimate loads for a series of beams with varying shear connexion, it can be seen that if the CP 117 full-connexion design is halved, the ultimate load will be reduced only slightly, and there will still be an adequate margin of safety against shear failure, which is believed to be more sudden than flexural failure. The remaining questions are therefore on the performance in relation to deflexion, connector loads and bending stresses. These, together with the reduction in ultimate moment, will now be discussed below under separate headings.

3.3.2 *Effect of partial-connexion design on ultimate loads*

The reduction in the ultimate moment due to the reduction in the number of connectors is relatively easy to calculate. Assuming that each connector develops its maximum shear force at failure, the maximum compression in the slab (C) can be calculated. The neutral axis (figure 3.6 is applicable here) can then be determined using the condition $C = T$, and the ultimate moment calculated accordingly. The result of a typical example is shown in figure 3.15.

When the shear connexion is reduced below the critical value, connexion failure takes over, and the reduced ultimate-moment curve (portion AB in figure 3.15) becomes irrelevant. Since complicated analyses are required to locate the critical shear connexion and calculate the failure load associated with shear failure, the ultimate-moment approach in figure 3.15 is insufficient to generalize a design method. To illustrate the effect of partial-connexion design on failure for various typical situations, Johnson's interaction diagram is reproduced in figure 3.16. This shows the relation between the partial-connexion moment (i.e. moment reached at failure, including shear failure) and the partial-connexion design expressed as a ratio n (which equals unity for full-connexion design).

The curve for uniformly loaded beams was included because shear failure was more likely to occur in this group of beams, as seen by the shifting of the critical-connexion point to the right. Similarly, the phenomenon of strain hardening is depicted because its presence in the steel tends to increase the tensile force (and hence the compressive force) in the slab ($C = T$) before flexural failure. Johnson also produced a similar diagram based on experimental results and came to the following conclusions.[6]

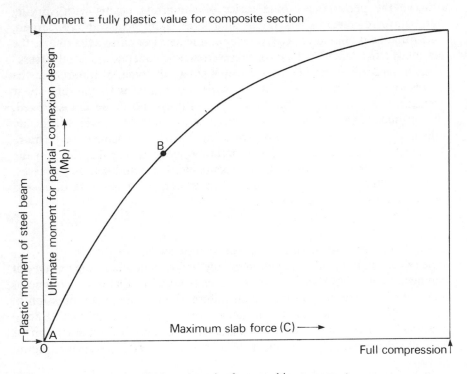

Figure 3.15   Effects of available compressive force on ultimate moment

Theoretical analysis and tests on beams both show that in partial-connexion design it is not possible to ensure that flexural failure always precedes shear failure. This does not matter, provided that the bending moment at which shear failure occurs is well above the design ultimate moment. The loss of strength of connectors at large slip is less sudden than was here assumed (particularly for welded studs of 19 mm and larger diameters) and in tests, longitudinal shear failure may be no more sudden than a flexural failure.

Johnson's results from all the beams with stud connectors confirm that the line AB in figure 3.16 represents a safe estimate of the flexural strength of beams with $N/N_f \geqslant 0.5$.

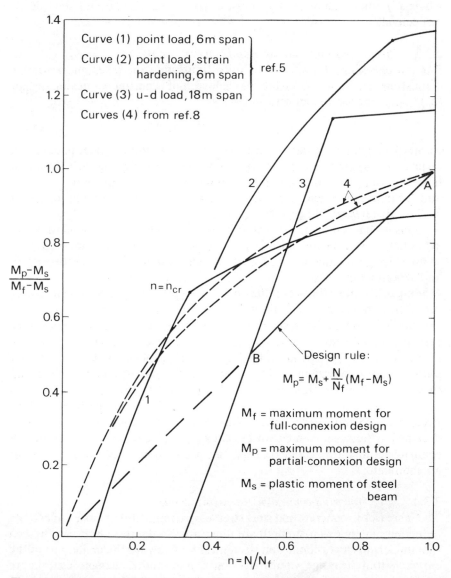

Figure 3.16 'Variation of flexural strength with connector ratio

It should be noted that the following equation can also be used to find the appropriate shear connexion to attain a given ultimate moment (M_p):

$$N = N_f \times \frac{M_p - M_s}{M_f - M_s} \tag{3.29}$$

where f, p and s denote full-connexion, partial-connexion and steel beam respectively.

3.3.3 *Effect of partial-connexion design on deflexion*

For the checking of the serviceability limit state of deflexion, a similar expression can be used to relate the deflexion of a partial-connexion design (δ_p) to the number of connectors used (N).

$$\delta_p = \delta_f + g(\delta_s - \delta_f)(1 - N/N_f) \tag{3.30}$$

From this equation, it can be seen that if $g = 1$, δ_p will be equal to δ_s and δ_f respectively when N is set equal to 0 and N_f respectively. Figure 3.17 shows, however, that $g = 0.5$ is a better approximation for the region between $n = 0.5$ and 1. Since values below 0.5 for the ratio are not recommended, equation 3.30 is satisfactory with $g = 0.5$.

Johnson has studied the validity of the above recommendation by examining the deflexions of a large number of beams. The study covered various design parameters, and made use of a range of existing theoretical and experimental data.

Some codes allow the use of limiting span-depth ratios as deemed to satisfy the deflexion criteria (i.e. there is thus no need to do any deflexion calculation). Since the recommended ratios given in most codes refer primarily to full-connexion design, the question arises as to whether the same ratios are valid for partial-connexion design. Although it is obvious that partial-connexion designs result in larger deflexions, it should be noted that they are in fact over-sized beams in relation to the ultimate load they are designed to sustain (see section 3.3.1). This problem has been studied by Frodin, Taylor and Stark,[9] who calculated the deflexions of over 1000 beams designed to the recommended ratios. They concluded that the same allowable span to depth ratios could be used for partial-connexion design without amendment.

3.3.4 *Effect of partial-connexion design on stresses*

The increases in concrete and steel stresses (extreme fibre stresses) due to the reduction of connexion strength are illustrated in figure 3.18. It can be seen that the sensitivity to connexion strength is very low for the major part of the curve. Within the range where n lies between 0.5 and 1, changes in strain are negligible. It is therefore quite reasonable to use the section modulus

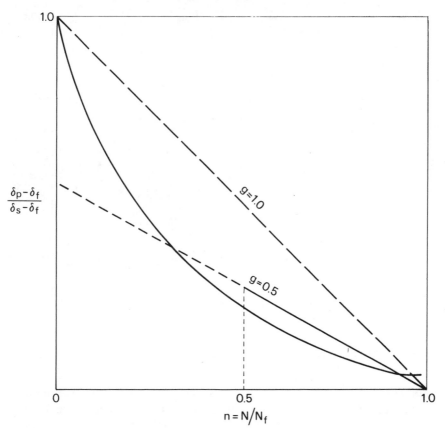

Figure 3.17   Effect of connector ratio on deflexion

approach for the calculation of stresses under working load conditions, as is the case with full-connexion design.

3.3.5   *Effects of partial-connexion on connector forces*

In the case of full-connexion design, longitudinal shear at the interface is calculated to ensure that the forces on connectors under working loads are not excessive. The elastic theory based on the transformed composite section is used, assuming the concrete slab to be uncracked and unreinforced. There is no need to use variable effective widths, so that equation 3.16 is applicable. For the partial-connexion design, a simple method can be used assuming that the shear calculated by the above method (q) remains unchanged for the same external load, in spite of the loss of interaction. If the number of connectors is then reduced to half of the full-connexion value ($n = 0.5$), then

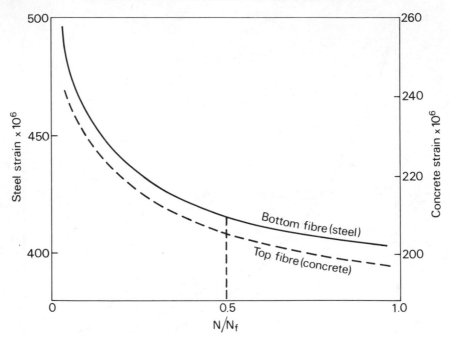

Figure 3.18   Effect of connector ratio on strains (point load at midspan)

it follows that each connector would be transmitting twice the load in order to maintain the same shear per unit length. Hence the equation for the connector force of the partial-connexion design is simply

$$Q_p = \frac{N_f}{N} Q_f = \frac{Q_f}{n} \tag{3.31}$$

3.3.6   *Useful expressions for the assessment of partial-interaction effects under working loads*

As far as design is concerned, codes allow the use of the method of section modulus for the calculation associated with the serviceability limit state. However, there may be occasions when a more accurate assessment of the partial-interaction effects is desirable. Although the rigorous solution of partial-interaction equations involves complicated mathematics, it is nevertheless possible to derive expressions (from equation 3.24) which are surprisingly simple to apply. Use may be made of magnification and reduction factors for deflexion and interface shear respectively due to partial interaction.

For a point load at midspan:
Magnification factor for central deflexion is

$$1 + 3\alpha(\theta - \tanh \theta)/\theta^3$$

Reduction factor for shear on end connector is

$$1 - 1/\cosh \theta$$

For uniformly distributed loads:
Magnification factor for central deflexion is

$$1 + \frac{24\alpha}{5} \times \frac{1}{\theta^2}\left(\frac{1}{2} - \frac{\cosh \theta - 1}{\theta^2 \cosh \theta}\right)$$

Reduction factor for shear on end connector is

$$1 - \tanh \theta/\theta$$

where

$$\theta = wl = \frac{(D+d)l}{2}\sqrt{\frac{K(1+\alpha)}{a\alpha(E_cI_c + E_sI_s)}}$$

4   Continuous Composite Beams

4.1   General design problems

Reverse bending occurs over the support of a continuous construction or at the beam-column joint of a framework. As far as ultimate capacity is concerned, continuous construction makes effective use of the reserve in strength due to structural redundancy, but incurs additional reinforcing steel for reverse bending and the labour required to place it. Some design problems, in addition to those associated with simply supported beams, are inevitable but not formidable. They arise partly because of continuity, and partly because of the change in the properties of the composite section due to reverse bending.

For the assessment of elastic behaviour, the beams have to be regarded as indeterminate structures in the calculation of bending moments. Unlike its steel and reinforced concrete counterparts, a composite section displays a significant difference in flexural stiffness in respect of positive bending (by sagging moment) and negative bending (by hogging moment). To establish a more accurate distribution of moments along the spans, the assumption of uniform stiffness can no longer be made without modification. The use of varying stiffness (or different stiffnesses constant over given lengths of the beam) is too complicated for design, even under simple loading and assuming complete interaction. The method described below and accepted by most Codes of Practice, consists of an analysis based on uniform stiffness followed by an adjustment of the bending moment diagram (redistribution) in accordance with some prescribed numerical rules.

The assessment of the ultimate load capacity is much simpler, since the bending moment diagram is determined chiefly by the plastic moments of the various sections, and therefore stiffness need not be considered. However, the designer should be aware that ideal plastic moments can be developed to a satisfactory extent only if certain ductility requirements are met. Under such ideal conditions, the section subjected to a bending moment equal to its fully plastic value should be able to sustain further deformation without significant reduction in its capacity (due to excessive crushing of concrete or buckling of the flange of the steel beam). With ductility, the load can be increased to develop further "plastic hinges", so that the use of fully plastic moments is valid at collapse. In spite of the above consideration of the history of loading, the design methods turn out to be quite simple. However, it is essential that the phenomenon is well understood. The underlying principles will be expounded in section 4.3.4.

Since the shear distribution along the span is substantially different from that of a simply supported beam, the slip distribution and hence also the effect of loss of interaction will have to be re-examined. Interface shear and design of shear connexion will be dealt with in section 4.4. Other problems common to both simply supported and continuous beams merit fairly detailed discussion and will be considered in chapter 6. These problems are: stresses due to the method and sequence of construction, effects of temperature, creep and shrinkage, and cracking of the concrete slab over the support.

4.2 Elastic behaviour of continuous composite beams

4.2.1 *Complete interaction*

As in the case of simply supported beams, complete interaction can be assumed in assessing the elastic behaviour, because it offers the only practicable method which gives results sufficiently accurate for the purpose of simple design. Even with this simplifying assumption, standard formulae for continuous beams with constant stiffness are inapplicable when cracking in the hogging moment region results in a substantial reduction in stiffness. Most Codes of Practice require the consideration of reduced stiffness in calculating the distribution of bending moments. The stiffness of the cracked region will then be based on the steel-beam, together with the slab reinforcement, contribution by the concrete being ignored.

To illustrate the various methods of analysis and design, an example of the symmetrical two-span continuous beam with point load at midspan[10] is used (figure 4.1). This example is chosen because further reference to it will be made in relation to some relevant experimental results. If cracking can be ignored so that constant stiffness can be assumed, the ratio of midspan moment to support moment is $m = 0.83$ (m being the positive value of the ratio). The length of the beam subjected to hogging moment can then be found.

To take account of reduction in stiffness in the hogging moment region, the use of the two constant stiffnesses as illustrated by the figure will be satisfactory, since any error over the tail-end of the bending moment diagram

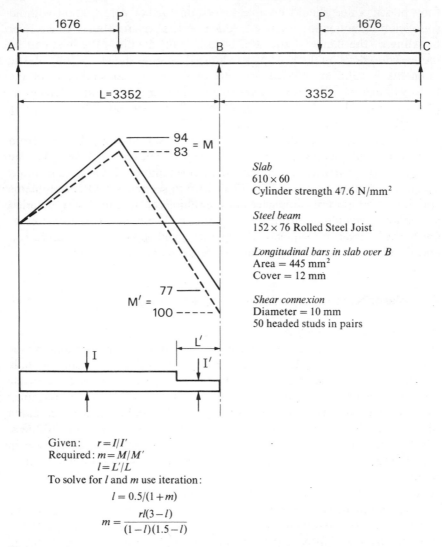

Slab
610 × 60
Cylinder strength 47.6 N/mm²

Steel beam
152 × 76 Rolled Steel Joist

Longitudinal bars in slab over B
Area = 445 mm²
Cover = 12 mm

Shear connexion
Diameter = 10 mm
50 headed studs in pairs

Given: $r = I/I'$
Required: $m = M/M'$
          $l = L'/L$
To solve for l and m use iteration:

$$l = 0.5/(1+m)$$

$$m = \frac{rl(3-l)}{(1-l)(1.5-l)}$$

Figure 4.1   Two-span continuous composite beam and distributions of elastic moment

will be small. The condition that m gives the correct solution is that the deflexion based on the assumed value of m should be zero at the end support, when integration proceeds from the internal support.

The value of m can be obtained by solving a cubic equation or from a plot of r against m (figure 4.2). The results are shown in figure 4.1. Comparing the reduced-stiffness solution (cracked) with the constant-stiffness solution

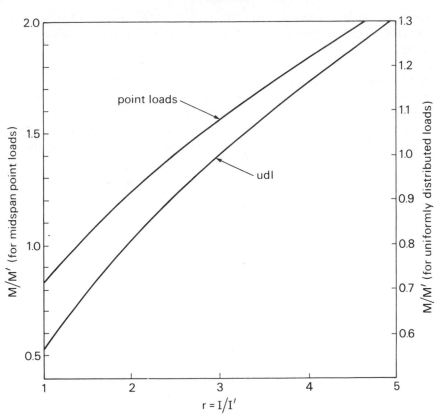

Figure 4.2 Graphs for distribution of moment along 2-span beams

(uncracked), it can be seen that the hogging moment in the cracked condition is lower by 23%. Because of the following equilibrium condition, the midspan moment is increased by 13% accordingly:

$$\frac{PL}{4} = M + \frac{M'}{2} \tag{4.1}$$

This phenomenon of redistribution of moment due to cracking is made use of in simplifying the design procedure. A given percentage of redistribution (about 20%) related to certain conditions is specified in Codes of Practice. The designer then uses the constant-stiffness approach, followed by a subsequent redistribution of moment from the support to the spans. It is worth noting here that there will be a further redistribution of moment from the support to both spans in the inelastic phase as plastic deformation develops in the hogging moment region.[11] A higher percentage of redistribution from the support is therefore appropriate for the ultimate limit state.

4.2.2 Incomplete interaction

In the discussion on simply supported beams, the effect of incomplete interaction on the elastic behaviour has been shown to be negligible for design purposes. In fact, complete interaction can be assumed, even with partial-connexion design, thus simplifying the procedure of checking the

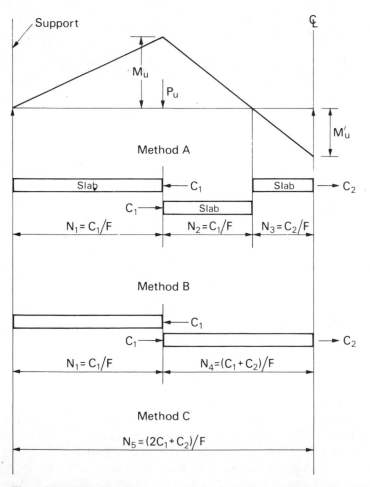

Figure 4.3 Methods of shear connexion design. Redrawn and adapted from Yam, L. C. P. and Chapman, J. C., "The inelastic behaviour of continuous composite beams of steel and concrete", *Proc. ICE*, Vol. 53, No. 7551, December 1972, pp. 487–501

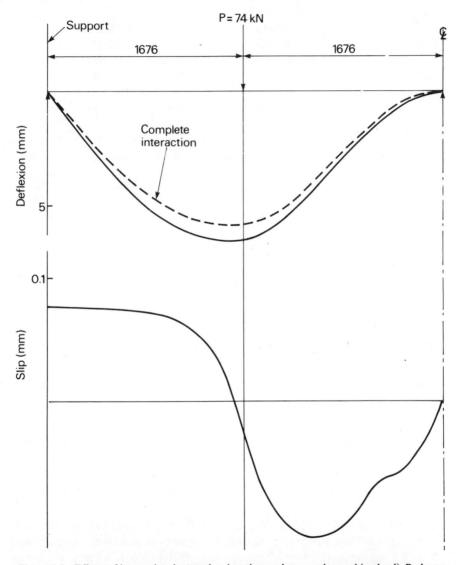

Figure 4.4 Effects of incomplete interaction (continuous beam under working load). Redrawn and adapted from Yam, L. C. P. and Chapman, J. C., "The inelastic behaviour of continuous composite beams of steel and concrete", *Proc. ICE*, Vol. 53, No. 7551, December 1972, pp. 487–501

various serviceability limit states. To illustrate the effects on continuous beams, the previous example (figure 4.1) will be used, the shear connexion designed in accordance with method B in figure 4.3.

The effect of incomplete interaction on the stiffness of the continuous beam can be seen from figure 4.4, which compares the deflected shapes under working load, obtained by assuming complete and incomplete interactions respectively. The difference in maximum deflexion is only 10%.

Similarly, the stress distributions are compared for the section under the point load and that over the central support (figure 4.5). It can be seen that,

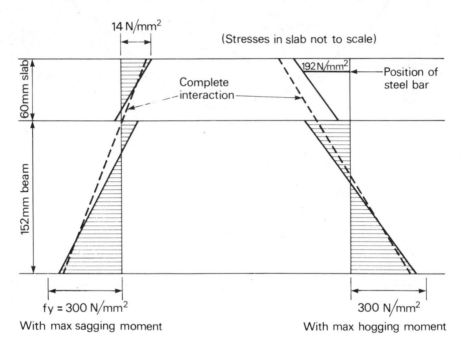

Figure 4.5   Stress distributions at working load. Redrawn and adapted from Yam, L. C. P. and Chapman, J. C., "The inelastic behaviour of continuous composite beams of steel and concrete", *Proc. ICE*, Vol. 53, No. 7551, December 1972, pp. 487–501

for the section under sagging moment, the effect of incomplete interaction on the extreme-fibre stresses is negligible. However, for the section under hogging moment, the loss of interaction results in a relatively noticeable reduction in the tensile strain at the top of the slab. It is interesting to note here that the loss of interaction is in fact beneficial, in that it alleviates the cracking of the concrete.

The slip distribution at working load is also shown in figure 4.4. It can be seen that the distribution is quite even (maximum values of positive and negative slips are about equal). The average force per connector is about 25% of the ultimate capacity. For the connector with maximum slip, the force is 45% of the ultimate capacity.

The assumption of complete interaction is therefore satisfactory in assessing the elastic behaviour of continuous beams. Of the various discrepancies, only the overestimation of tensile strain at an internal support is of any significance.

4.3 Inelastic behaviour of continuous composite beams

4.3.1 *Complete interaction*
The effect of incomplete interaction on the inelastic behaviour of continuous composite beams will be discussed in section 4.3.5. Methods of shear connexion design will also be given to ensure that premature failure of the shear connexion is reasonably remote. When adequate shear connexion is provided, and local failures such as buckling of the compression flange of the steel beam at the support and longitudinal splitting of the slabs are prevented by appropriate provisions, the development of plastic hinges at collapse may be assumed for the calculation of ultimate loads. Whether the ideal values of the fully plastic moment can be used, depends on the extent of plastic deformation at the hinges at the point of collapse.

For the hinge subjected to hogging moment, the effective section consists of the steel beam and the longitudinal reinforcement in the slab. This all-steel section can be assumed to have sufficient ductility to maintain the fully plastic moment as hinge rotation takes place. For the hinge subjected to sagging moment, however, the effect of crushing on the fully plastic moment has to be examined.

4.3.2 *Moment-curvature relation for a composite section*
For a steel beam with an I-section subjected to pure bending in one plane, a typical moment-curvature relation is as shown in figure 4.6(a) where strain-hardening of the steel is neglected. The ideal moment-curvature relation assumed for the simple plastic theory of steel structures is shown in the same figure. This ideal relation has been found to be satisfactory in predicting collapse loads of most steel structures and some reinforced concrete structures. Some implications of these idealizations are:

(1) When the bending moment at a section reaches the fully plastic moment (M_p), the curvature can become infinitely large, so that a finite change of slope can occur over an infinitely small length of the member. A hinge can therefore be imagined to be inserted in the member at this section, which permits rotation (in the same direction as the applied moment) of any magnitude while the moment remains constant at the value of M_p.
(2) The fully plastic moments for bending in both senses (i.e. sagging and hogging) are equal.
(3) Regardless of the bending history of the section, it can always attain the fully plastic moment of magnitude M_p for bending in either sense.

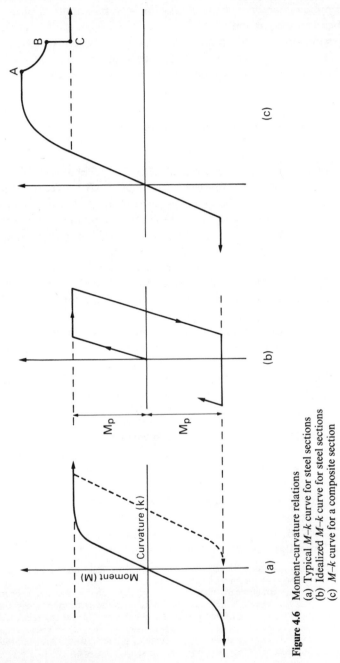

Figure 4.6 Moment-curvature relations
(a) Typical M–k curve for steel sections
(b) Idealized M–k curve for steel sections
(c) M–k curve for a composite section

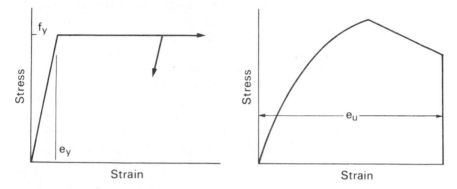

Figure 4.7 Idealized stress–strain curves for steel and concrete

A similar moment-curvature curve for a typical composite section is shown in figure 4.6(c). Complete interaction is assumed, and the stress-strain relations for steel and concrete are as shown in figure 4.7, tensile strength of concrete being ignored. Crushing of the concrete slab starts at A (i.e. the maximum concrete strain reaches e_u) while the I-beam has completely yielded in tension. As the curvature increases, more concrete crushes and the moment of resistance is reduced. At B, the stiffness of the concrete slab is reduced to such an extent that the composite section becomes unstable—continued crushing of concrete occurs at the same curvature until, at C, the bare steel beam alone contributes to the moment of resistance. The post-crushing behaviour (portion AC) is due to a reduction in both the concrete force and the moment arm. The phenomenon of instability occurs (at B) when the negative stiffness of the concrete slab exceeds the positive stiffness of the steel beam. The condition of instability is discussed in Reference 12 and is represented by the equation

$$\frac{\partial C}{\partial n} = \frac{\partial T}{\partial n} \qquad (4.2)$$

This equation indicates that a lowering of the neutral axis while keeping the same curvature will reduce C and T by the same amount. Instability therefore occurs under this condition, since there will be more than one state of equilibrium. The quantitative aspect of the moment-curvature relation need not be taken into account in design, but it is important to recognize the phenomenon which distinguishes the behaviour of composite beams from that of reinforced concrete beams as far as plastic design is concerned. The behaviour of showing a reduction in moment with increasing curvature is known as *strain-softening* as a contrast to strain-hardening.

Another characteristic in relation to the bending of a composite section is the irreversibility as regards attaining the fully plastic moments when bent successively in both senses. In the first place, any crushing of the concrete amounts to a reduction in the effective depth of the slab, so that the sagging ultimate moment cannot be reached subsequently. Secondly, although hogging hinge rotation can be tolerated while a constant bending moment is maintained, the cracking of the slab may have impaired the strength of the concrete, so that the sagging ultimate moment of resistance will not be attained subsequently. Hence the idealization as regards reversibility of bending assumed in the plastic theory would not be valid for composite sections.

4.3.3 Some problems in plastic theory of continuous composite beams

To illustrate a few problems which will occur when the plastic theory is applied to continuous composite beams, the two-span continuous beam (figure 4.1) is used as an example. A ductile moment-curvature relation is first assumed (figure 4.8) and the plastic moments for bending in two senses are unequal. Applying the simple plastic theory, the collapse load corresponding to the mechanism shown in the figure is $P = 149$ kN. The beam behaviour from zero load to collapse is described below:

(1) Elastic analysis shows that the moments at B and D increase linearly ($M_B/M_D = 1.2$) until the moment at B reaches the fully plastic value of 51 kN m, when the load is $P = 88$ kN.

(2) As the load increases, the section at B behaves as a hinge, and the beam is then equivalent to two simply supported beams of spans AB and BC. The additional load on each span required to bring the moments at D and E to the fully plastic values is given by

$$2(89\ 000 - 42\ 000) \div 1524 = 61 \text{ kN}$$

Hence, the total load on each span at collapse is

$$88 + 61 = 149 \text{ kN}$$

The simple plastic theory is therefore applicable for the above method of loading. To illustrate the breakdown of the plastic theory when the question of rotational capacity arises, a further stage of loading is considered below.

(3) At the end of (2), suppose the plastic moments D and E have just been reached but not exceeded. When the load at E is gradually removed, the beam will behave as if the load Q of 149 is gradually applied at E in the opposite direction. Since the application of Q will decrease the sagging curvature at E and increase that at D, a hinge will form at D while unloading occurs in the span BC, as shown in the figure. If the hinge at D can rotate at constant moment, the moment distribution due to Q can be obtained by treating BC as simply supported. The resulting bending-moment distribution after the complete removal of the load at E is also shown in the figure.

This equilibrium condition is possible only if there is sufficient ductility as implied by the moment-curvature relation in figure 4.8. However, if the strain-softening character of figure 4.6(c) is taken into account, the hinge at D may not maintain a constant moment of resistance, and failure may occur before the complete removal of the load.

Another mode of loading which may invalidate the plastic theory is the application of a single point load at D. It can be shown that the first hinge

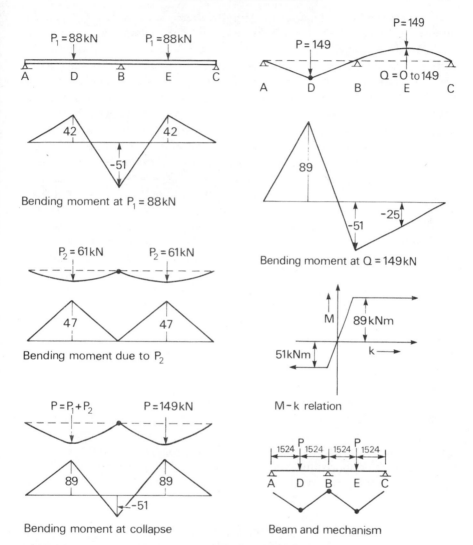

Figure 4.8 Effects of method of loading on plastic collapse

forms at D while P is 143 kN. Since any increment of P will increase the curvature at D, whether the fully plastic moment can be attained at B depends on the ductility of hinge D.

4.3.4 *Plastic design of continuous composite beams*
In spite of the above departure from the ideal fully-plastic behaviour, the simple plastic theory is still applicable to the design of continuous composite

beams, provided that some further simple criteria are satisfied. It has been shown that the main distinguishable features of the composite sections are the limited rotational capacity and the vulnerability in respect of bending in both senses. The latter feature does diminish the generality of the plastic theory but poses no major problem as far as design is concerned. The considerations of critical loading cases and moment envelopes will indicate the extent to which a section may be subjected to bending in both senses, and appropriate provisions such as detailing should be made to provide the necessary resistance. Rotational capacity therefore constitutes the main problem, assuming that verifications against flange buckling and longitudinal splitting of the slab have been made.

Three solutions to this problem will be discussed below. In the first place, if the strain-softening hinge is the last hinge to complete the collapse mechanism, then the rotational capacity, though limited, will not affect the magnitude of the collapse load. Secondly, if the strain-softening hinge is developed before collapse, some design parameters may be adjusted to maximize the rotational capacity of the hinge. Thirdly, if the above two cases do not apply, then it may be assumed, for the purpose of calculating a collapse load, that collapse occurs as soon as the strain-softening hinge develops. This approach is conservative because there will be critical sections over which the plastic moments have only partially developed, and it is these lower values that have to be used for the calculation of the collapse load.

Strain-softening hinge as the last hinge

If the strain-softening hinge is the last to form, intermediate hinges are by implication ideal plastic hinges. This renders the study of the sequence of hinge formation relatively easy, because the progressive reduction in the degree of redundancy simplifies the calculation of bending moments. Thus, the elastic bending-moment diagram is used to locate the first hinge (the load at which this occurs is hereafter referred to as the "single-hinge load"). Then follows another calculation of elastic bending moment for a higher load, now acting on a beam with an additional hinge (this higher load will be called "two-hinge load"). Thus, elastic analysis can continue to be applied to portions of the continuous beam between hinges until the collapse mechanism is formed and the corresponding collapse load obtained.

For two-span beams with equal point load at midspan, Yam and Chapman[10] have shown that the simple plastic theory is applicable for a wide range of design parameters. They studied experimental and theoretical results of a series of beams to analyse the effects of varying areas of longitudinal reinforcement, depth of slab and material properties. it was concluded that the use of the plastic theory was satisfactory for the symmetrical two-span continuous beams, loaded by midspan loads or uniformly. Subsequent tests by Hamada[13] and by Mallick and Chatto-padhyay[14] confirmed the validity in respect of beams with point loads.

Rotational capacity
Examples in which strain-softening hinges develop first are now discussed, again using an elastic analysis to study the onset of hinge formation. Figure 4.9 illustrates two typical three-span continuous beams tested by Mallick and Chattopadhyay, who examined the performance of strain-softening hinges

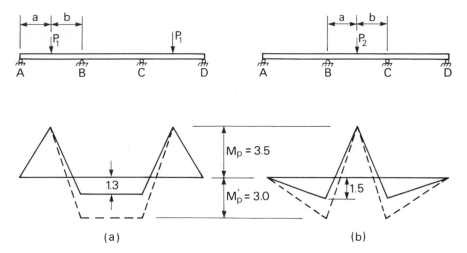

Figure 4.9 Comparison between elastic moment distribution and ideal distribution for plastic collapse

subjected to high curvature in a series of 18 beams. For equal spans with constant flexural stiffness, the following formulae for elastic moments under unsymmetrical loading can be used:

For P_1 acting in span AB only

$$M_B = 4P_1 ab(L+a)/15L^2 \tag{4.3}$$

$$M_C = P_1 ab(L+a)/15L^2 \tag{4.4}$$

For P_2 acting in span BC

$$M_B = P_2 ab(2L+5b)/15L^2 \tag{4.5}$$

$$M_C = P_2 ab(7L-5b)/15L^2 \tag{4.6}$$

For the bending moment under the point load, the relevant expression can be obtained by considering equilibrium of the loaded span with known moments at the end supports. The inelastic behaviour can be effectively assessed by constructing two bending-moment diagrams. In figure 4.9, dotted lines are used to represent the ideal plastic moment distribution (which depends on two values only: $M_p = 3.5$, $M_p' = 3.0$). The elastic moments are then calculated for unit load and multiplied by such a factor as to allow one section to reach the plastic moment (i.e. the single-hinge load is found and the corresponding elastic bending-moment diagram constructed). This implies that the elastic load is increased until a hinge begins to form. The result is shown in full lines in the figure.

These diagrams illustrate two convincing examples of "brittle mechanisms", that is, plastic mechanisms with strain-softening hinges. In case (a) for example, the potential strain-hardening hinge is at less than half its full strength ($1.3 \div 3.0 = 0.43$) when the strain-softening hinge has developed. To develop the ideal collapse load, therefore, the sagging hinge must undergo considerable rotation without significant loss of resistance. The same condition of adequate ductility is required for the beam in case (b). From the series of tests on such "brittle mechanisms", Mallick and Chattopadhyay[14] observed the ability of the strain-softening hinges to sustain large deformations. Actual collapse loads were found to be very close to the values based on the simple plastic theory. Of the 18 beams tested, five were loaded as in figure 4.9(a) and were more critical in respect of rotational capacity required at collapse. The ideal plastic moment at the support sections was not fully developed at collapse, the largest reduction being 20%, although the reduction in collapse load was only 9%.

In spite of some encouraging results from research work such as that described above, existing knowledge on three-span composite beams is still insufficient to justify a general application of the simple plastic theory without restrictions. Thus, the above tests all involved composite sections with the plastic neutral axis within the slab for sagging bending, which in fact belonged to a family of ductile sections. Furthermore, if lengths of span are very different, simultaneous formation of hinges is far from being attainable. For example, the beams in figure 4.10 would have an ideal moment distribution if the span ratio is about 0.75 (assuming all fully plastic moments are about equal). For span ratios outside the range indicated in the figure, the strain-softening hinge would be at less than half its fully plastic value when the other similar hinge had just developed.

There is very little guidance in Codes of Practice or elsewhere on the applicability of the plastic theory to indeterminate composite beams. Engineering judgment has to be exercised. An effective method of assessing the collapse behaviour consists of a comparison of the plastic moment

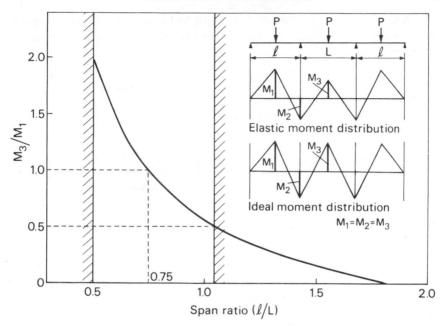

Figure 4.10 Effects of span ratio on attainment of ideal moment distribution

diagram with the elastic moment diagram corresponding to the single-hinge load, together with a consideration of "section ductility". A discussion on the attributes of ductility is given below.

Rotter and Ansourian[15] have carried out a systematic study of ductility characteristics of composite sections subjected to bending by sagging moments. They analysed a wide range of cross-sections and formulated some qualitative and quantitative criteria of ductility. The numerical criterion is fairly simple and useful, but inevitably has limitations. Strain-hardening of the steel beam is assumed, and the strain at which this occurs has to be known, whereas the extent of strain-hardening is neglected in the assessment. According to this criterion, ductility will be satisfactory if the following is satisfied:

$$A_s f_y < k_1 k_3 B (D+d) f_c' \left(\frac{e_u}{e_u + e_{sh}} \right) \tag{4.7}$$

where $k_1 k_3 = 0.72$ when the neutral axis is in the slab and when the slab is not reinforced (or has equal amount of reinforcement above and below the neutral axis)

f_c' = cylinder strength of concrete

e_{sh} = strain at start of strain-hardening

e_u = ultimate strain in concrete.

The qualitative guidelines are fairly simple to follow, but are based on the presence of strain-hardening in the steel beams. They are summarized below (see figure 4.11):

Slab thickness increases ductility because it lengthens the partially plastic regime BC.

Slab width increases ductility because it allows a greater strain-hardening gain in moment (DE).

Strain-hardening of steel increases ductility.

Higher steel yield stress reduces ductility.

Weaker concrete reduces ductility,

Reinforcement placed in a single layer near the top of the slab improves ductility, because it decreases the depth of concrete stress block and so increases the curvature at crushing.

Geometry of the steel beam section has negligible effect on ductility.

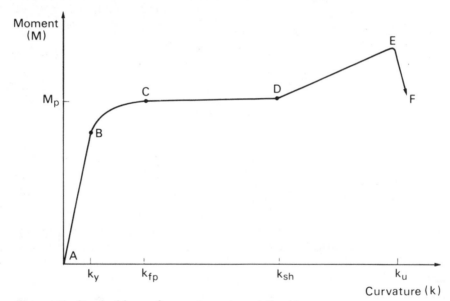

Figure 4.11 Standard forms of moment-curvature relationship

Partial plastic design

If a comparison of the plastic moment diagram and the single-hinge moment diagram shows that the first hinge is the strain-softening type while the other potential hinges are reasonably near their ideal plastic values, then the single-hinge load can be taken as the collapse load to avoid further analysis. This "partial plastic design" offers a simple method of checking the ultimate limit state in practical design. As has been pointed out in section 3.3.1 on partial-connexion design, beam sizes are often governed by other considerations than ultimate resistance, and limit-state calculation will then consist of checking that no collapse occurs at a given level of loads.

When the moment-curvature relations for all the relevant hinges are available, it is possible to obtain a more accurate and also higher partial plastic load. Although calculation of this kind would not be appropriate for design, it is shown here (figure 4.12) mainly to illustrate features which contribute to high collapse loads.

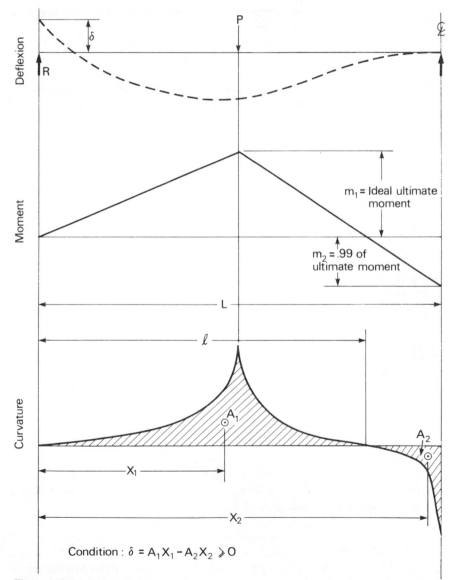

Condition : $\delta = A_1 X_1 - A_2 X_2 \geqslant 0$

Figure 4.12 Criterion of plastic mechanisms

For the symmetrical two-span beam in the figure, if m_2 is the support moment at the single-hinge load, then for compatibility, the deflexion at the end support should vanish:

$$\delta = A_1 X_1 - A_2 X_2 = 0 \tag{4.8}$$

where A_1, A_2 = areas of curvature diagrams derived from the bending moment diagram

X_1, X_2 = distances from the end support of the corresponding centroids of the areas A_1 and A_2.

If δ is positive, P must be increased to obtain a better solution; hence positive δ implies a safe solution.

It should also be noted that additional load on the span would "spread" the bending moment diagram and increase A_1, δ and hence the collapse load. In general, spreading load on a span tends to improve ductility of the beam.

To study the applicability of the simple plastic theory to the design of continuous composite beams with two equal spans, Yam and Chapman[10] carried out a systematic analysis of collapse loads for a wide range of beam parameters and loading configurations. For symmetrical loading consisting of midspan point loads and uniformly distributed loads, the plastic theory is found applicable. For other cases of loading which are likely to induce "brittle mechanisms", the concept of partial plastic design was adopted to study the extent to which the support moment may fall below its ultimate value at the onset of crushing at the point of maximum sagging moment.

For the symmetrical system shown in figure 4.13, the magnitude of the central support moment is reduced as the loads are placed nearer the end supports. However, the ideal ultimate load is only slightly reduced (less than 2%) because the contribution of the hogging moment to the ultimate load becomes less significant when the loads are near the supports.

When unequal midspan loads are applied (figure 4.14), reduction in central support moment also occurs. The greatest reduction occurs when one span is unloaded, there then being a 10% reduction in the ideal ultimate load.

It thus appears that the plastic theory is applicable for all cases of symmetrical loading investigated, while a 10% reduction in the ideal collapse load would be appropriate for non-symmetrical loads. The 10% allowance is on the safe side, as is confirmed by experimental study on eight beams tested by Mallick and Chattopadhyay.

Plastic design methods in codes of practice
The qualitative conditions for the simple plastic theory to be applicable to continuous composite beams are summarized below. Apart from the

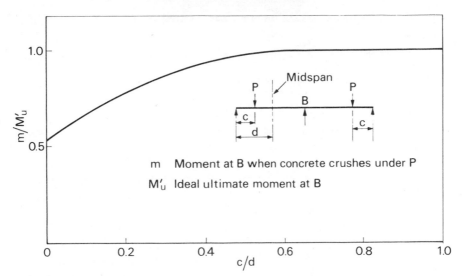

Figure 4.13 Effect of position of loads on reduction of support moment. Redrawn from Yam, L. C. P. and Chapman, J. C., "The inelastic behaviour of continuous composite beams of steel and concrete", *Proc. ICE*, Vol. 53, No. 7551, December 1972, pp. 487–501

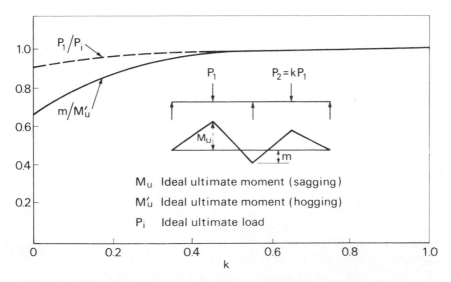

Figure 4.14 Effect of unequal loading on reduction of support moment. Redrawn from Yam, L. C. P. and Chapman, J. C., "The inelastic behaviour of continuous composite beams of steel and concrete", *Proc. ICE*, Vol. 53, No. 7551, December 1972, pp. 487–501

provision against local buckling and longitudinal splitting of the slab, they are:

(a) At the single-hinge load, each potential hinge should have developed a high proportion of its fully plastic moment.
(b) Strain-softening hinges should have adequate ductility.
(c) Load within a span should be well spread over the sagging moment region.

Because it is not yet possible to quantify the above conditions or formulate deemed-to-satisfy rules to cover the wide range of variables, most Codes of Practice give very little guidance on plastic design and tend to be restricted in some of their rules. In respect of the sequence of hinge development under (a), no satisfactory guidance is given. Redistribution of moment from a sagging hinge to a support is not allowed (support-to-span redistribution is allowed, with percentage of redistribution classified). Rules on the ratio of spans are given to ensure that spans are about equal, so that potential hinges do carry substantial loads at the single-hinge load (figure 4.10). In respect of rotational capacity under (b), both qualitative and quantitative guidance are absent. In respect of load spreading under (c), it is stipulated that if a specified portion of any span is supporting more than half of the total load on that span, then the load on that portion would be too concentrated to allow the use of the simple plastic theory.[16]

4.3.5 Incomplete interaction

In the previous discussions on simply supported beams (section 3.1.2) it has been shown that (if adequate shear connexion is provided) the loss of interaction due to slip has negligible effect on the ultimate moment of resistance of the composite section. For continuous beams, the effect of slip on ultimate moment in respect of both sagging and hogging bending is of interest, and an illustration of the stress distributions at failure across the composite section will be particularly useful. Figure 4.15 shows these distributions, which are analytical results on symmetrical two-span beams subjected to midspan point load and uniformly distributed load respectively. Stress distributions based on complete interaction are also shown (as dotted lines). It can be seen that the loss of interaction has little effect on the stress distributions and negligible effect on the ultimate moments. As in simply supported beams, the shear and hence the slip under a point load have higher gradients, and therefore the strain difference at the interface is greater. This explains the relatively high degree of interaction for the section of maximum sagging moment of the uniformly loaded beam.

It has also been pointed out, during the discussion of the elastic behaviour of continuous composite beams (section 4.2.2) that the loss of interaction alleviates the cracking of the slab in the hogging moment region. This stiffening effect has been confirmed by observations on continuous beams

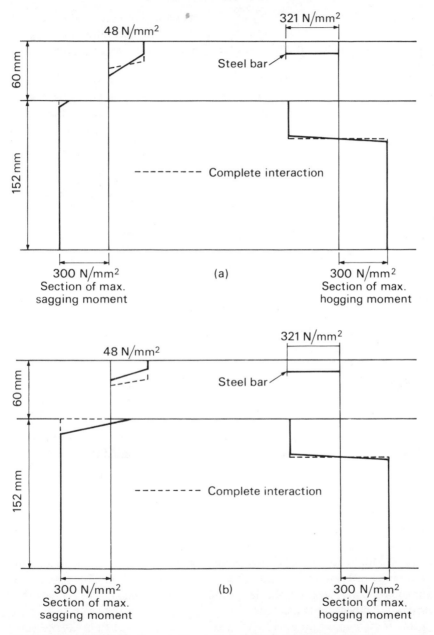

Figure 4.15 Stress distributions at ultimate load: (a) UDL (b) point load. Redrawn from Yam, L. C. P. and Chapman, J. C., "The inelastic behaviour of continuous composite beams of steel and concrete", *Proc. ICE*, Vol. 53, No. 7551, December 1972, pp. 487–501

tested in the laboratory. Cracking over the hogging moment region occurred at loads much higher than those based on cracked sections, and higher stiffness was reflected in the unusually high ratio of hogging moment to sagging moment.

Figure 4.16 shows the hogging moment region of a two-span beam near collapse (at 87% of ultimate load). As a result of the interface slip, tensile strains in the slab were lower and more even. No cracking was visible at this load, and the region with tensile strains exceeding 100 microstrains is shown in the figure.

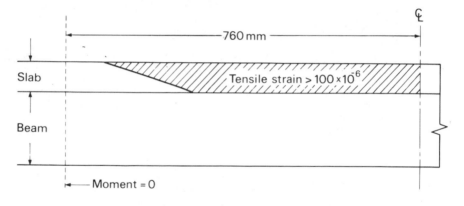

Figure 4.16   Hogging moment region of continuous beam near collapse

Evans and Kong[17] have made detailed microscopic investigations of the extensibility and microcracking of the *in situ* concrete in composite beams with precast prestressed concrete members. They concluded that

microscopic cracks always developed when the strain reached about 100 microstrains and that they had no significant influence on the general strength of a concrete structure.

From figure 4.16, the amount of uncracked concrete can be seen to be quite appreciable. Furthermore, the shaded portion will also contribute some amount of stiffness.

It is important to note that this stiffening effect at the hogging moment region tends to bring the neutral axis up towards the slab. This leads to a higher compressive strain in the flange of the steel beam and thus renders the beam more vulnerable to local instability.

Another incomplete interaction effect worth noting is a shift in the position of maximum strain from the section of maximum moment when the beam is loaded uniformly (see figure 4.17). This discrepancy between the locations of maximum strain and maximum moment may have little influence on the design calculations. It should nevertheless be considered in detailing the slab reinforcement of a continuous composite beam when the loading is fairly uniform.

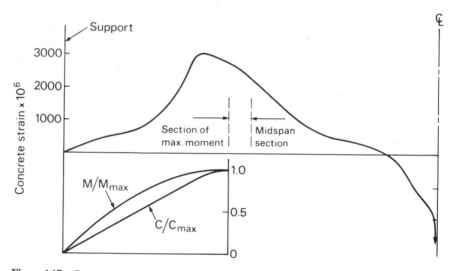

Figure 4.17   Concrete strain along beam at failure (UDL)

4.4   Design of shear connexion in continuous composite beams

As in simply supported beams, the number of shear connectors is related to the force in the slab required to develop the ultimate moment. Because of hogging bending in a continuous beam, connectors are required to induce tension as well as compression. These forces, together with the relevant slab elements on which they act, are shown in figure 4.3. The compressive force is the compression in the slab for a fully plastic section, as in simply supported beams. The tensile force is taken as that due to the yielding of the longitudinal reinforcement only, since cracked concrete is assumed to have no tensile strength.

Having determined the number of connectors, there appear to be three sensible methods for spacing them (figure 4.3):

Method A: Consider each slab element between zero and maximum force and space connectors uniformly for the slab element in question. (The figure shows three elements of this type.)

Method B: Where the section with zero force is not at the support, redefine slab element as slab between sections of maximum compression and maximum tension; then follow Method A.

Method C: Use uniform spacing throughout the continuous beam.

Differences in beam behaviour between Methods A and C can be observed from the tests on two quarter-scale continuous beams by Teraszkiewicz[18] (CB 1 based on Method A and CB 2 based on C). With overall uniform spacing, CB 2 had shear connexion failure. However, there were signs of spalling indicating that flexural failure was imminent. The ideal ultimate load was exceeded mainly because of strain-hardening in the steel beam. The slip distribution at 87% of ultimate load, shown in figure 4.18, does not appear to be quite uniform. In fact, the average force per connector at shear

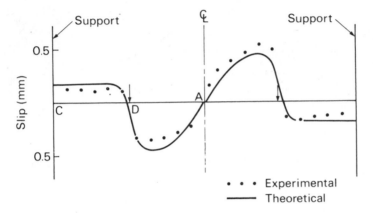

Figure 4.18   Slip distribution near failure (CB 2 with uniform connector spacing). Redrawn from Yam, L. C. P. and Chapman, J. C., "The inelastic behaviour of continuous composite beams of steel and concrete", *Proc. ICE*, Vol. 53, No. 7551, December 1972, pp. 487–501

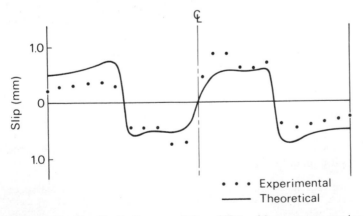

Figure 4.19   Slip distribution near failure (CB 1 with connector spacing by Method A). Redrawn from Yam, L. C. P. and Chapman, J. C., "The inelastic behaviour of continuous composite beams of steel and concrete", *Proc. ICE*, Vol. 53, No. 7551, December 1972, pp. 487–501

failure was 90% of the ultimate capacity for that portion with negative slip (DA), and was only 78% for that portion with positive slip (CD). With three different spacings, but the same total number of connectors, CB 1 had flexural failure and the ideal ultimate load was again exceeded. The slip distribution at 94% of ultimate load (figure 4.19) shows that the maximum positive and negative slips were closer than in CB 2. The average forces per connector for positive slip and negative slip were 93% and 89% respectively.

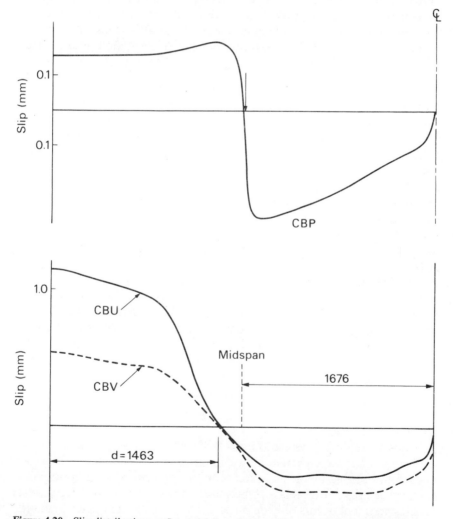

Figure 4.20 Slip distributions at flexural failure. Redrawn from Yam, L. C. P. and Chapman, J. C., "The inelastic behaviour of continuous composite beams of steel and concrete", *Proc. ICE*, Vol. 53, No. 7551, December 1972, pp. 487–501

Thus this method of connector spacing is more efficient than the overall uniform spacing employed in CB 2.

For completeness, the slip distribution corresponding to Method B (designated CBP for continuous beam with point load) is shown in figure 4.20, which consists of analytical results taking account of inelasticity and incomplete interaction. The ideal ultimate load is again attained, with flexural failure. The slip distribution at flexural failure is shown in the figure and is reasonably even. The average force per connector is 58% and the maximum connector force 76% of the ultimate capacity. This shows that spacing connectors uniformly between points of maximum and minimum moment is satisfactory in preventing premature shear failure under static load.

Uniformly distributed load (UDL)
Shear connexion design for UDL involves no additional requirement, except that the position of maximum sagging moment at collapse should be calculated from the corresponding bending moment diagram. To illustrate the difference in the beam behaviour between the procedure recommended above and the assumption that maximum moment occurs at midspan, two beams with connexions designed by these two methods are analysed (CBU based on maximum moment at midspan and hence connexion identical to that of CBP; CBV based on moment diagram at collapse—Method B used for both). Beam CBU had connexion failure while CBV has flexural failure, although the ultimate loads are nearly equal. The slip distributions are compared in figure 4.20. The reason why shear failure occurs in CBU but not in the case of simply supported beams is that in CBU the maximum compressive concrete force occurs away from the midspan (figure 4.17). The position of the maximum moment should therefore be considered in designing the shear connexion. By constructing the bending moment diagram so that the maximum and minimum moments are respectively equal to the fully plastic moments, the position of maximum moment can be found (figure 4.21).

Off-centre point load
Previous examples show that Method B of connexion design which provides uniformly spaced connectors between sections of maximum sagging and maximum hogging moment is satisfactory in preventing connexion failure. It might be argued that this is so because in the examples considered so far, hogging moment occurs only on a short length of the span, and so another example is studied. For CBO, in which the point load is near the end support, a greater proportion of the span is subjected to hogging bending moment.

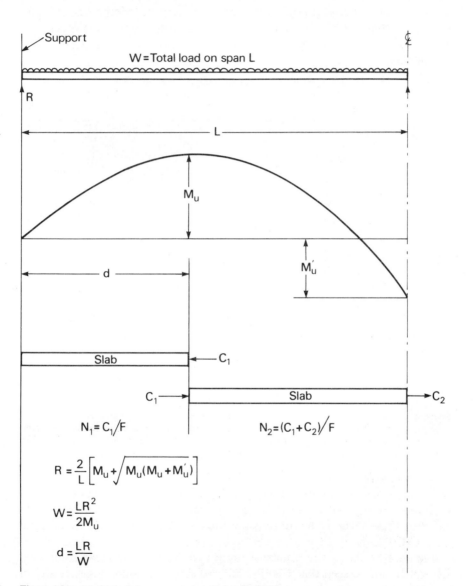

Figure 4.21 Method of shear connexion design for UDL

The shear connexion in CBO 1 is designed on the maximum-minimum moment basis (Method B) while that in CBO 2 is designed by considering sections of maximum (or minimum) and zero moment (Method A). Figure 4.22 shows that CBO 2 gives a lower maximum negative slip. However, the increase in ultimate moment is only 1% and Method B may be regarded as satisfactory.

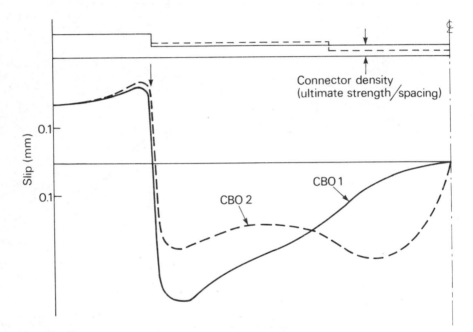

Figure 4.22   Slip distribution in CBO 1 and CBO 2 at ultimate load. Redrawn from Yam, L. C. P. and Chapman, J. C., "The inelastic behaviour of continuous composite beams of steel and concrete", *Proc. ICE*, Vol. 53, No. 7551, December 1972, pp. 487–501

Two point loads on span

When there are concentrated heavy point loads between sections of maximum and zero moment, it is advisable not to space the connectors uniformly between these sections, but to consider the shear diagram at plastic collapse as in CBPP 2.

Figure 4.23 shows the results of these two methods of shear connexion design. It can be seen that CBPP 1 has a larger maximum negative slip at crushing failure. The same stipulation for connector spacing as in simply supported beams (figure 3.9) should be followed when there are heavy concentrated loads between the sections of maximum (or minimum) and zero moment.

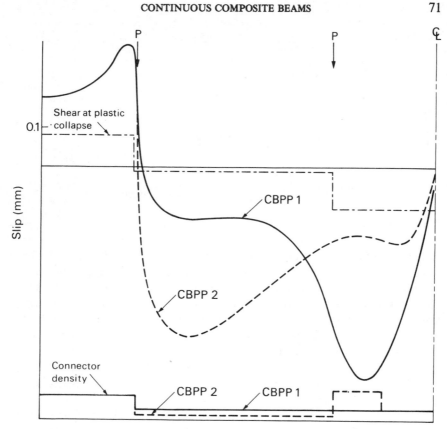

Figure 4.23 Slip distribution in CBPP 1 and CBPP 2 at ultimate load. Redrawn from Yam, L. C. P. and Chapman, J. C., "The inelastic behaviour of continuous composite beams of steel and concrete", *Proc. ICE*, Vol. 53, No. 7551, December 1972, pp. 487–501

5   Shear Connectors

5.1   Introduction

The performance of composite structures depends on an effective transfer of shear stress at the interface. Tests on composite T-beams have shown that the natural bond alone is insufficient to ensure satisfactory interaction at high loads, and that the mode of collapse initiated by bond failure is somewhat catastrophic. The use of mechanical shear connectors is therefore essential for the design of composite beams, unless the steel section is fully encased in concrete and adverse loading is excluded. Furthermore, the presence of connectors enhances the shear resistance of the natural bond at low loads, which leads to the paradox that "shear connectors are unnecessary as long as they are there".

Shear connectors also serve the function of holding the slab down onto the steel beam. Although torsional forces and loads acting downwards on the steel beam are obvious causes of vertical separation at the interface, there are other causes due to the non-linearity of either material. These causes cannot easily be described in terms of a physical phenomenon but could be appreciated if it is noted that the uplift force is related to the second derivative of the slab moment along the interface. Suppose a portion of a composite beam is subjected to a constant bending moment. Because of slip and yielding, the ratio of the slab moment to the moment of the steel beam may not remain constant along the length. The slab moment can therefore vary along the length and the "curvature of slab moment" thus causes uplift in spite of a constant external moment. Uplift forces have been observed in many tests on simply supported composite T-beams. All connectors in common use are therefore specially shaped to provide adequate resistance to uplift as well as to slip. In normal design, calculation related to uplift is not required.

When moving concentrated loads are considered, as in the design of bridge decks under traffic loading, the shear near the midspan could give rise to a major design problem. From the influence-line diagram for a point load it can be seen that a section at or near midspan is subjected to a reversal in vertical shear as the point load passes that section. Similarly, the interface shear and hence the connector force fluctuate as the bridge deck is loaded by moving vehicles. In these circumstances, fatigue failure becomes the design criterion, and shear connectors must be provided and designed to specific loading spectra.

The structural performance requirements of shear connectors can therefore be summarized as shear resistance, the ability to tie down and fatigue strength. The question pertinent to design is how to evaluate a connector in relation to these three performance requirements. Since the structural behaviour of connectors is complicated and not amenable to simple calculations, a section is given below to discuss the factors affecting the performance of connectors, following a qualitative treatment of their structural behaviour. But before proceeding further, the common types of connector are described so as to provide physical models for subsequent reference.

5.2 Types of shear connector

There are many types of shear connector in use in Europe and North America, but inevitably the choice in any country is limited by local experience in design and construction. In particular, values of static and fatigue strengths of connectors recommended by Codes of Practice are normally based on extensive research and testing. Current British Codes give information on some commonly used types of connector, which are shown in figure 5.1.

Shear connector types are broadly divided into two categories—rigid connectors (bars with hoops and tees with hoops) and flexible connectors (channels and studs). In addition to the obvious difference in stiffness (load per unit slip), these two groups differ also in the mode of failure. Rigid connectors tend to cause higher stress concentrations in the surrounding concrete, resulting in crushing or shearing failure of concrete, or even weld failure. The failure mode of a flexible connector is more consistent and less catastrophic.

The headed stud is the most widely used type of connector in composite construction. Its popularity lies mainly in the economies resulting from the speed of the stud welding process. The welding is carried out by "firing" the stud onto the steelwork from a portable hand-tool connected via a control unit to mains or to a generator. Highly skilled operatives are not required and welding of the stud takes 10–15 seconds. The semi-automatic process also

(a) Stud connector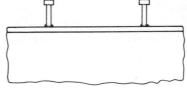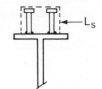

(b) Bar connector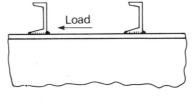

(c) Channel connector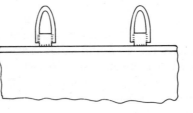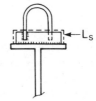

(d) Tee connector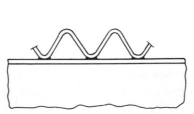

(e) Helical connector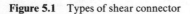

Figure 5.1   Types of shear connector

results in uniformity of quality and involves only very low heat input to the parent metal.

The use of epoxy resin adhesive to stick the concrete slab to the steel beam has been studied and some success claimed. Apart from the requirement of site control (temperature and surface cleanliness) not readily available, there are doubts on the performance in relation to uplift and fatigue life. Further study appears to be necessary before recommendation for its use can be made.

5.3 Connector behaviour and factors affecting performance

The qualitative behaviour of connectors in respect of horizontal shear, uplift and repeated loading will be discussed here to provide a basis for the understanding of the various factors which can affect their performance. The basic tests for this assessment of connector behaviour are the push-out test for static loads and fatigue test for repeated loads.

Push-out tests on connectors
The specimen consists of a short length of steel beam connected to two small *in situ* slabs as shown in figure 5.2. The natural bond at the interface is prevented from occurring by greasing the steel flanges before casting the slab (except of course in the case of high-strength friction-grip bolts). The slabs are bedded in mortar on the lower platen of a testing machine, and compression is applied at the top cover plate of the vertical specimen, through a steel ball to ensure concentric loading. Standardized parameters related to push-out tests are given in current Codes of Practice.

The test can enable the full load-slip curve to be obtained, but for design purposes the main interest is in the ultimate capacity. It is also useful to check that the uplift separation is not excessive, normally less than half the interface slip at the corresponding load level. Some load-slip curves are shown in figure 5.5.

It should be noted that the load-slip curve for an individual connector in a composite beam may be different from that obtained from a push-out test. The overall beam behaviour and the slip distribution, however, can be satisfactorily depicted by the use of results from push-out tests. Connector strength based on push-out tests is conservative, considering that the concrete in a beam is normally under a higher compression.

Before discussing the factors affecting the resistance of connectors to horizontal shear, it is appropriate to examine the mode of failure,[19] which is illustrated in figure 5.3 together with the stress distribution on the shank of the headed stud. The figure shows the presence of stress concentration near

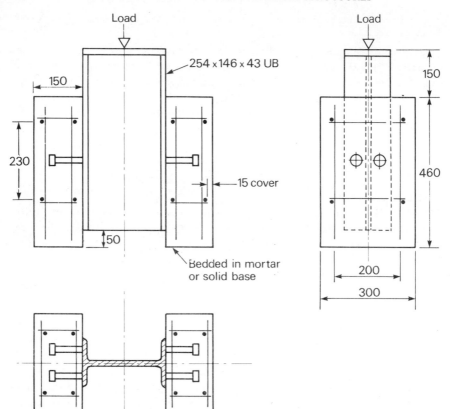

Figure 5.2   Push-out test on shear connectors. (All dimensions are in millimetres. Reinforcement should be of 10 mm diameter mild steel.) Redrawn from BS 5400: "Steel, concrete and composite bridges, Part 5: Code of Practice for Design of Composite Bridges", British Standards Institution 1979

the base of the stud. High stresses are possible here because the concrete is restrained by the steel flange, the connector and the reinforcement. As expected, the two major modes of failure are crushing of the concrete surrounding the connector (for studs with large diameters) and connectors shearing off at the base (for slender studs).

Intuitively, connector strength appears to be related to concrete strength, yield stress of the stud material and the degree of concrete containment. Tests by Menzies[20] and at Imperial College, London have established a definite increase in connector strength with higher concrete strength, as shown by figure 5.4. When lightweight aggregate concrete is used, there is a noticeable reduction in connector capacity (for headed studs, the reduction

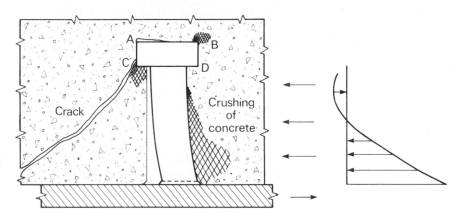

Figure 5.3 Mode of failure of a shear connector

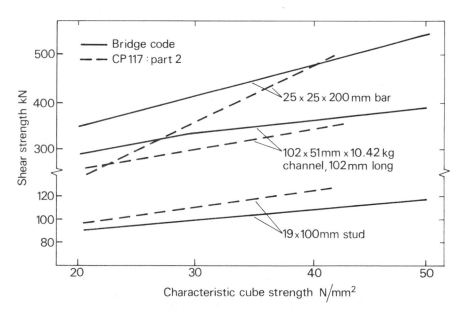

Figure 5.4 Effects of concrete strength on connector capacity

can be taken as 15%, provided that the density of the lightweight concrete exceeds 1400 kg/m³). There is also a correlation between connector strength and the properties of the connector itself. Thus connector strength of a headed type is roughly proportional to the square of its diameter and, for studs embedded in strong concrete, their strength depends on the ultimate tensile strength of the stud material (but not the yield strength). Shorter studs provide small bearing areas against concrete and have lower strengths, but when the length exceeds about four times the shank diameter, the effect of increasing the length becomes negligible. There is no systematic research work on the relation between connector strength and the degree of containment of the concrete, but most test results have indicated positive correlations. Thus the use of ribbed plates for the top flange has been found to increase not only the bond, but also the connector strength significantly. The general performance of connectors is found to be better when the slab is in compression than in tension (as in the case of bending under hogging moment), but test results have shown that compression in the slab increases connector stiffness more than its strength.

Tests have also shown that connector behaviour is substantially different in haunched slabs where lateral restraints are low, and that beams with haunches tend to fail suddenly. Connector strength, however, would not be affected if the sides of the haunch lie outside a line drawn at 45° from the outside edge of the connector. Furthermore, this edge must have sufficient concrete cover as illustrated in figure 5.6.

Although the strength of the connector is the only parameter used in the design calculations, it is instructive to understand the implications of some key parameters implied in its load-slip curve. Figure 5.5 shows some load-slip curves from which three further parameters can be noted: initial stiffness (say up to 50% of full strength), overall stiffness and ductility. Initial stiffness affects only the magnitude of the interface slip under the static working load and is the least significant of the three parameters. The importance of the overall stiffness can be appreciated if the effect of connector flexibility on the ultimate load of a beam is considered. Suppose five beams with the same flexural strength are designed using five different types of connector characterized by curves A to E in Figure 5.5. Although the connectors in each of these beams should develop the same force at connexion failure, the beam with connexion E will develop the lowest force at flexural failure. However, the reduction in ultimate load due to this overall flexibility is significant only in special circumstances. A study by Yam and Chapman[5] has shown that a critical combination is a concentrated point load on a beam with lightweight concrete, the ultimate load falling below the CP 117 value by 10%. Beam failure can also be affected by connector ductility, or limiting slip at connector failure. According to the same study, a uniformly loaded long-span beam with low-ductility connectors could have connexion failure if the number of connectors provided is about 20% below that recommended by CP 117.

Research work on the resistance of connectors to uplift forces is limited but, on the other hand, there are no major problems caused by uplift, except in such structures as composite box girder bridges in which there is local

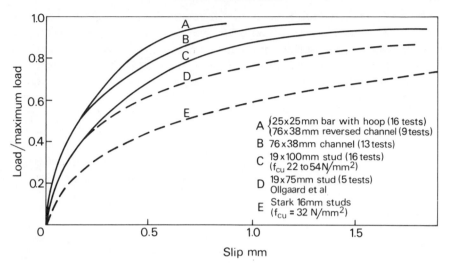

Figure 5.5 Load-slip curves for stud, channel and bar connectors

distortion of the flange due to buckling and diagonal tension in the webs. Assessment of connector performance in respect of uplift is based on the assumption that the holding down is effected by the top of the connector. Thus, sufficient area should be provided by the underside of the head of the stud, the lower surface of the top of the hoop (bar connector) and the lower face of the top flange of a channel connector. Furthermore, the concrete around this holding down force should be effective in resisting the tension thus created. It follows that the top part of the connector should ideally be embedded in the compression zone of the slab. Current Codes also recommend that both the connector length and its projection above the bottom transverse reinforcement should be adequate.

The fatigue life of a connector depends chiefly on the connector-to-beam weld, and the mode of failure can be fracture of the steel flange or connector shearing off. Since the area of weld is normally related to the cross-sectional area of the connector, current codes place a limit on the stud diameter to avoid excessive weakening of the steel flange. Tests by Osborne-Moss[21] have shown that the fatigue life of headed studs embedded in concrete in tension is less than that of identically loaded connectors embedded in compressed concrete.

5.4 Transverse reinforcement

Shear connexion design for the ultimate limit state is based on connector failure at maximum interface shear. Another mode of failure is shear failure of the concrete surrounding the connector, producing a "tunnel" as illustrated by the dotted lines in Figure 5.6(a). The part equivalent to the tunnel lining is known as the shear surface or shear plane. In addition to the provision of connectors to resist the interface shear, a further check is therefore required to ensure that the maximum interface shear does not exceed the resistance provided by the shear planes. As seen from the figure, this shear resistance is due to the concrete and the transverse reinforcement. When transverse bending can be neglected, the shear resistance on the shear planes is the sum of two components:

(a) Product of the area of the shear plane and some factored shear strength of the concrete. For unit length of beam, the area of the shear plane is equal to the cross-sectional length of the shear plane (L_s), which is the length of the dotted line in figure 5.6. The Bridge Code recommends 0.9 N/mm² for the shear strength for normal-density concrete; hence this component is $0.9 L_s$.

(b) Product of the area of transverse reinforcement crossing the shear planes and some factored yield strength of the reinforcement. For a unit length of beam, the area of transverse reinforcement A_e is given in figure 5.6 and the factored strength is 0.7 of the nominal yield strength (f_{ry}); hence the component is $0.7 A_e f_{ry}$.

The effectiveness of the transverse reinforcement in resisting shear depends on the state of stress of the reinforcement and the surrounding concrete. Two unfavourable conditions are: tension in the bottom portion of the slab due to transverse bending (sagging moment), tension near the top of a connector (such as underside of the head of a stud) due to connector holding down the slab. Consequently, effective area (A_e) is defined in such a way (below, together with the table in figure 5.6) as to exclude, where appropriate, the transverse reinforcement provided in flexure. Furthermore, when transverse bending produces tension in the bottom of the slab, the above two components for the shear resistance on plane 2–2 have to be reduced. On the other hand, the beneficial effect of transverse bending by a hogging moment can be made use of by increasing the shear resistance. The appropriate formulae for both of these effects are given in the Bridge Code. To take account of the influence by the uplift force, bars are assumed to be effective only if they are placed at a clear distance of at least 40 mm below the level where uplift forces are applied (underside of top part of a connector).

Given below are definitions of cross-sectional areas of transverse reinforcement (per unit length of beam) to assist the understanding of the table in figure 5.6. These definitions apply only to bars which are fully anchored on both sides of the shear plane.

A_t = reinforcement placed near the top of the slab forming the flange of the composite beam and may include that provided for flexure.

A_b = reinforcement placed in the bottom of the slab or haunch at a clear distance not greater than 50 mm from the nearest surface of the steel beam, and at a clear distance of not less than 40 mm below that surface of each shear connector that resists uplift forces, including that bottom reinforcement provided for flexure.

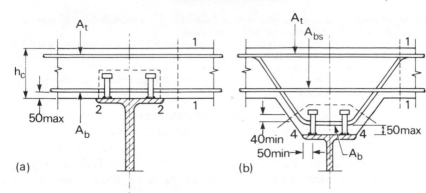

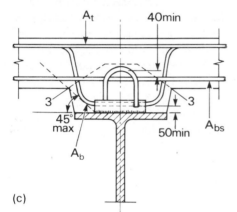

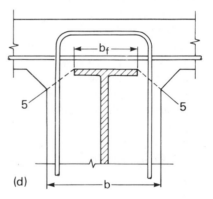

Shear plane type	A_e
1–1	$(A_t + A_b)$ or $(A_t + A_{bs})$
2–2	$2A_b$
3–3	$2(A_b + A_{bs})$
4–4	$2A_b$
5–5	Cross-sectional area of stirrups (both legs) crossing the shear plane

Note: For shear plane type 5–5 L_s = total length of shear plane minus one third b_f.

Figure 5.6 Shear planes and transverse reinforcement. Redrawn from BS 5400: "Steel, concrete and composite bridges, Part 5: Code of Practice for Design of Composite Bridges", British Standards Institution 1979

A_{bs} = other reinforcement in the bottom of the slab placed at a clear distance greater than 50 mm from the nearest surface of the steel beam.

A_{bv} = reinforcement placed in the bottom of the slab or haunch, but excluding that provided for flexure, which complies in all other respects with the definition of A_b above. It should exceed 0.5 A_e in haunched beams.

A_e = the reinforcement crossing the shear planes assumed to be effective in resisting shear failure along the planes.

It is sometimes necessary to provide additional transverse reinforcement in the bottom of the slab to prevent longitudinal splitting around the connectors. The simple criterion can be taken as the proximity of the shear plane to the connectors, which is measured by the ratio of L_s (2–2 in figure 5.6) to the slab thickness. If this ratio does not exceed 2, the additional area provided should at least be $0.8 \times$ (slab thickness)/f_{ry}. The additional reinforcement is not required if there is a sufficiently large compressive force acting normal to the shear planes. The limiting value for this compression per unit length of beam may be taken as $1.4 \times$ the slab thickness (in N/mm^2).

There is also a limit on the shear resistance provided by the shear planes, which is $0.15 L_s f_{cu}$ with the proviso that the cube strength lies between 20 and 45 N/mm^2. For further information on detailing of transverse reinforcement (including curtailment and minimum area), its spacing near the end of the beam and the relevant factors for lightweight aggregate concrete, reference should be made to the relevant Codes of Practice.

5.5   Composite floors with profiled steel sheeting

5.5.1   *Introduction*
Slabs cast on corrugated steel decking are an interesting form of composite construction in which the steel deck acts also as a shear connexion. Composite floors of this type have been used for a number of years in North America and have recently gained much acceptance in Europe. They are mainly used in buildings, primarily for floors but occasionally for roofs, to provide a permanent formwork which also functions as bottom reinforcement. The corrugations provide convenient ducts for the incorporation of services within the floor. Figure 5.7 illustrates various types of floor with profiled sheets.

5.5.2   *Composite action*
Although the profiled steel sheet can be used merely as a permanent shuttering, composite action must be considered to achieve economy of design. Composite action can refer to the floor or the T-beam when a steel beam is used. In the latter case, the shear connexion is substantially similar to the normal composite construction involving mechanical shear connectors,

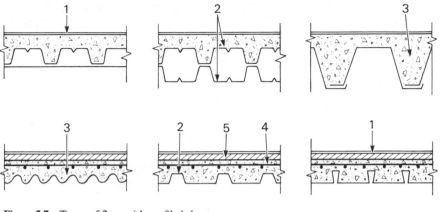

Figure 5.7 Types of floor with profiled sheets
     1. floor finish
     2. profiled sheet
     3. structural concrete
     4. mesh reinforcement
     5. topping

and the design will not be discussed here. For composite floors, no mechanical connexions are usually provided because, apart from the cost, welding of connectors to the very thin metal sheets is impracticable. Transfer of shear is achieved by one or more of the following provisions (see figure 5.8):

    (*a*) The profile shape
    (*b*) Indentation or embossments of the profile
    (*c*) Anchor straps
    (*d*) Holes in the corrugations of the profile
    (*e*) Bars welded to the profile
    (*f*) End anchorage (by headed studs)

It should be noted that uplift is prevented by many of the above provisions, such as the profile shapes and the inclining of the dimples to the vertical (in opposite directions on the two sides of a corrugation).

5.5.3 *Design*

There is a wide variety of profiled sheets and no general methods of design are available. Manufacturers normally provide all the relevant information and indeed make the design calculation much simpler. If further information is required, testing may have to be carried out; procedures recommended by the relevant Code of Practice should be followed. Only a brief outline of design procedure is given below.

Before composite action is acquired, the adequacy of the profiled sections should be assessed in respect of buckling of the compression flanges and

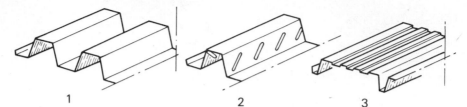

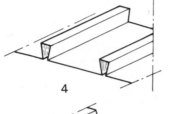

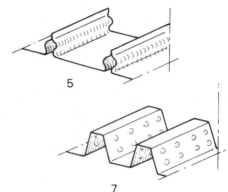

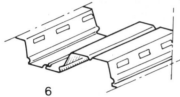

Figure 5.8   Types of profiled sheet
1. plain profiled sheet
2. lugs on the webs of ribs
3. stiffeners on the flanges
4. dovetailed troughs
5. shaped troughs
6. cusps on the webs and intermediate stiffeners
7. dimples

webs, moment capacity and excessive deflexions. The loads to be allowed for are the self-weight of the deck, the wet concrete and construction loads such as concreting plant and personnel. For the composite floor, the appropriate ultimate limit states to be considered are: flexural failure for spanning in the direction of the ribs, shear failure at the interface, and shear failure in concrete near the support. For serviceability, deflexion and cracking over any continuous support have to be considered. If mesh reinforcement is used, however, continuous slabs can be designed as simply supported.

Further detailed information on methods of calculation, together with measures against fire and corrosion, can be obtained from the relevant British Standards[22] and Handbooks[23].

Figure 5.9   A plate girder bridge with composite concrete deck, High Wycombe

Figure 5.10   A composite building under construction, Newcastle-upon-Tyne

6 Miscellaneous Problems

6.1 Shrinkage, creep and temperature effects

6.1.1 *Definition of the problems*

Concrete undergoes long-term deformations, irrespective of whether it is subjected to stresses. Shrinkage can affect unstressed concrete without temperature changes. When concrete is exposed to dry air (strictly speaking, unsaturated air), the water lost first is the "free water" held in the capillaries, which causes practically no shrinkage. It is only when the adsorbed water is removed as drying continues that the cement paste undergoes a volume change, thus causing shrinkage. Shrinkage also occurs owing to carbonation—when carbon dioxide present in the air reacts, in the presence of moisture, with hydrated cement minerals in the concrete. An idea of the orders of magnitude of the shrinkage-time relation can be obtained from figure 6.1, which also shows the influence of the type of aggregate.[24]

The increase of strain in concrete under a sustained stress is known as *creep*. Figure 6.2 shows the increase in strain of a concrete specimen loaded in dry air.[25] To simplify the presentation, a constant elastic strain due to load is assumed, but it should be noted that this strain depends on the rate of loading and that the modulus of elasticity of concrete increases with age. Furthermore, shrinkage and creep are not independent phenomena and are treated as a single parameter in most design calculations. Another way of describing creep is the progressive decrease in stress with time when concrete is subjected to constant strain. This concept forms the basis of the use of a reduced modulus of elasticity to take account of creep. Creep-time curves have very similar shape and the following guidelines[25] are helpful in the prediction of creep: they are for the usual range of structural concretes, loaded at ages of 28 and 90 days, and stored at a relative humidity of 50 to 100%.

18 to 35% (average 26%) of the 20-year creep occurs in 2 weeks.
40 to 70% (average 55%) of the 20-year creep occurs in 3 months.
64 to 83% (average 76%) of the 20-year creep occurs in 1 year.

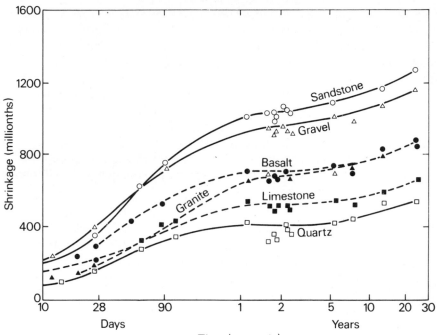

Figure 6.1 Shrinkage of concretes of fixed mix proportions but made with different aggregates. Redrawn from Troxell, G. E., Raphael, J. M. and Davis, R. E., "Long-time creep and shrinkage tests of plain and reinforced concrete", *Proc. ASTM*, Vol. 58, pp. 1101–1120, 1958

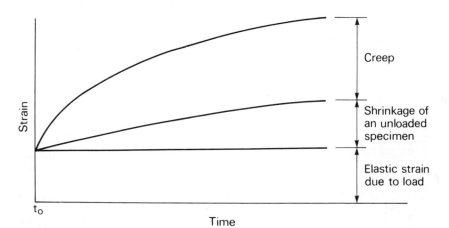

Figure 6.2 Change in strain of loaded and drying specimen. Adapted and redrawn from Neville, A. M., *Properties of Concrete*, Pitman Publishing, second (metric) edition 1973, reprinted 1975

Since the magnitudes of strains due to shrinkage and creep are of the same order as the elastic strain under working load, it is often necessary to take account of them in design calculations, especially for the limit state of deflexion. The effect of temperature, on the other hand, involves a problem of a slightly different nature. Owing to a non-uniform distribution of temperature across a section of a member, the corresponding differential thermal strains create local stresses as well as overall deformation. The non-uniform distribution can be caused by the different degrees of exposure to sunlight and wind, and the differences between the thermal capacity and conductivity of steel and concrete. Temperature effects constitute a major consideration in the design of bridges, especially in relation to the allowance for movement at joints and bearings.

6.1.2   Common basis of calculation

To formulate a common method of analysis for the above three actions, the concept of a free strain is used here for the strain at a point in the material which is free from stresses under these three actions. The slice-element in figure 6.3 is assumed to be unstressed initially (zero time as reference datum) and to have developed the following strains after a period of time:

$$e_f = e_s + e_c + e_t \qquad (6.1)$$

where e_s = strain due to shrinkage,
         e_c = compressive strain due to creep,
         e_t = strain due to decrease in temperature,
         e_f = total of the above strains, also the free strain corresponding to
                 zero stress.

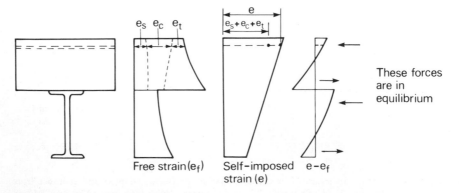

Figure 6.3   Deformation of a section to maintain equilibrium (without external load)

If the element is now subjected to a compressive strain of e (relative to the initial state), then the stress will not be proportional to this strain, but to the difference $e - e_f$:

$$\sigma = E(e - e_f) \qquad (6.2)$$

where E is the Young's Modulus for steel (E_s) or concrete (E_c) as appropriate.

If the free strain distribution is as shown in figure 6.3 after a given period of time measured from the initial state, then the initially unstressed section will become stressed if it is constrained against deformation. If the section is free to deform, the above equation indicates that a completely stress-free condition is possible only if the strain distribution e coincides with the distribution of free strain. Study related to the theory of elasticity shows that such a completely stress-free state is not compatible with the behaviour of elastic beams. Instead, if a section is free to deform, planes will remain plane, and the deformation can be defined by a uniform strain (due to axial compression in this example) together with a curvature due to pure bending about the neutral axis, provided that interface slip is negligible. (See the self-imposed strain diagram.) These two unknowns for the strain diagram can be solved using two conditions of equilibrium—zero axial force and zero bending moment respectively.

Thus, from equation 6.2, the equilibrium conditions for the compressive force and the bending moment (positive for sagging bending) are respectively

$$\int E(e - e_f)dA = 0$$
$$\int E(e - e_f)ydA = 0$$

or

$$\int EedA = \int Ee_fdA \qquad (6.3)$$
$$\int EeydA = \int Ee_fydA \qquad (6.4)$$

where dA = element of area,

$\quad\quad\quad y$ = distance of elemental area from the neutral axis.

The right-hand sides of the above two equations are the compression and bending moment respectively due to the free strain diagram only (say C_f and M_f). These equations are now re-written below to obtain the deformations required to restore equilibrium: a uniform axial strain at the neutral axis e_n and a curvature k.

$$(E_sA_s + E_cA_c)e_n = C_f$$

Hence

$$e_n = C_f/(E_sA_s + E_cA_c) \qquad (6.5)$$
$$k(E_sI_s + E_cI_c)(1 + \alpha) = M_f$$

Hence

$$k = \frac{M_f}{(E_sI_s + E_cI_c)(1+\alpha)} \qquad (6.6)$$

where equation 6.6 is based on the flexural stiffness of the composite section derived previously (see equations 3.11 and 3.12). Combining these equilibrating strain distributions with the free distribution, the resultant stress distribution can be obtained, as shown in figure 6.3.

The above analysis shows that, for an unrestrained beam, the effect of shrinkage, creep or temperature variation consists of the setting up of local stresses and an overall deformation. The latter results in a movement along the longitudinal axis, together with a rotation of the member by internal stresses alone. Most beams are designed to allow for a limited horizontal movement so that no further stresses would be induced. The curvature due to internal stresses, however, will cause further stresses in redundant members, such as continuous beams, in which forces are induced at the internal supports to restrain the beam from deflecting. Any additional stresses due to such restraints are known as *secondary effects*. The principle of consistent deformation can be conveniently applied to calculate these effects. If the internal supports of a continuous beam are removed and the deflexions of the resultant simply supported beam determined, the internal support reactions for the continuous beam can then be obtained by working out the forces required at the supports to bring the deflexions back to zero.

Equations 6.5 and 6.6 apply generally to strains due to shrinkage, creep and temperature with any distribution over the cross-section. To illustrate the application of the equations to a simple distribution, the free strain due to temperature assumed in CP 117 (Part 2: 1967) is used to derive expressions for stresses due to differential temperature between the concrete slab and the steel beam. The distribution consists of a constant strain e_1 over the slab depth and zero strain in the steel beam. Hence the compression and bending moment due to the free strain are obtained by integrating the right-hand sides of equations 6.3 and 6.4:

$$C_f = E_cA_ce_1 \qquad (6.7)$$

$$M_f = E_cA_ce_1(n - D/2) \qquad (6.8)$$

The slab force can be written down by considering the three components due to axial strain e_n, curvature k and free strain e_1:

$$C = E_cA_ce_n + kE_cA_c(n - D/2) - E_cA_ce_1 \qquad (6.9)$$

where the second term containing k and n refers to bending of the composite section according to equations 3.6 and 3.8. The values of e_n and k can be obtained from equations 6.5 and 6.6:

$$e_n = \frac{E_c A_c e_1}{E_c A_c + E_s A_s} \tag{6.10}$$

$$k = \frac{E_c A_c e_1 (n - D/2)}{(E_c I_c + E_s I_s)(1 + \alpha)} \tag{6.11}$$

Eliminating n between equations 6.11 and 3.8, simple expressions for the curvature and hence slab force of equation 6.9 can be derived for the purpose of design. They are equivalent to those in CP 117 (Part 2: 1967) and are given below:

$$k = \frac{\alpha e_1}{(1 + \alpha) d_c} \tag{6.12}$$

$$C = -\frac{E_c A_c \times E_s A_s}{E_c A_c + E_s A_s} \times \frac{e_1}{1 + \alpha} \tag{6.13}$$

where d_c is the distance between the centroid of the steel beam and the midplane of the slab, i.e. $(D+d)/2$ and α is calculated from equation 3.12. Equation 6.13 is primarily for the checking of forces on the end connectors. If fibre stresses are required to be checked, the above expressions can be used in the following manner:

stress at top of concrete slab $\sigma_1 = C/A_c + kE_c I_c/z_{c1}$ (6.14)

stress at bottom of concrete slab $\sigma_2 = C/A_c - kE_c I_c/z_{c2}$ (6.15)

stress at top of steel beam $\sigma_3 = -C/A_s + kE_s I_s/z_{s3}$ (6.16)

stress at bottom of steel beam $\sigma_4 = -C/A_s - kE_s I_s/z_{s4}$ (6.17)

where z_{c1}, z_{c2} are the section moduli for the top and bottom of the concrete slab alone, assumed to be uncracked, and z_{s3}, z_{s4} are the section moduli for the top and bottom of the steel beam alone.

The slab force due to differential temperature drops to zero at the free end and the relatively abrupt change produces interface shear of a magnitude sufficiently high to be taken into account in design (see figure 6.4). The solution for the shear along the beam has to take account of slip and is very similar to that shown in section 3.1.1. The variation of the slab force along the span due to free strain e_1 in the slab can be expressed by two equations (equivalent to equations 3.25 and 3.26):

$$\frac{d^2 C}{dx^2} - w^2 C = \frac{K e_1}{a} \tag{6.18}$$

$$C = \frac{K e_1}{a w^2} (\cosh wx - \tanh wl \sinh wx - 1) \tag{6.19}$$

Since equation 6.19 is too complicated for design, most codes recommend that a constant interface shear can be assumed to act over a given length of beam near the end support (one-fifth of effective span) and that the value of the shear can be taken as the slab force (equation 6.13) divided by the length of beam.

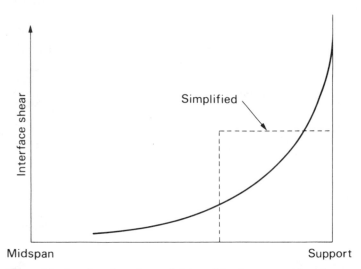

Figure 6.4 Interface shear due to shrinkage (simply supported beam)

6.1.3 *Design requirements*
Creep
The treatment of creep effects in British codes is relatively straightforward. The more elaborate approach adopted in Europe is mainly due to the more frequent use of prestressing and the aversion to any cracking of concrete. The simple approach consists of modifying the modulus of elasticity to take account of the increase in strains under sustained loading. This makes it no longer necessary to calculate the increase in strain separately, which can therefore be ignored in the remaining calculations. In fact, some deemed-to-satisfy provisions such as span-depth ratios in deflexion control have already taken account of the effects of creep. The calculation of the reduced modulus, or effective modulus, based on the creep coefficient ϕ is given by

$$E_c' = E_c/(1+\phi) \tag{6.20}$$

Values of the creep coefficient can be obtained from Appendix C of Part 4 of BS 5400. Part 5 of this Bridge Code also gives a simple table of reduction factors for the effective modulus where the concrete specification complies with certain limits.

Creep effects on columns, however, are less straightforward since the reduction in stiffness affects the ultimate capacity. Theoretical study suggests that the reduction in strength becomes appreciable as the concrete contribution factor exceeds about 0.3. Basu and Sommerville[31] derived magnification factors to be applied to that part of the loading which is considered as long-term. There are as yet insufficient tests on composite columns under long-term loading to enable definite design recommendations to be made. Furthermore, there is the practical difficulty of identifying the long-term portion of the total load. In buildings, for example, it is generally impossible to specify the type of use or the exact mode of loading valid throughout the lifetime of the structure. One practical method of allowing for creep effects in column design is the use of reduced strength of concrete, an idea which has been adopted in the Bridge Code.

Shrinkage and temperature
Shrinkage and temperature effects can be considered together, although there is the inconvenience caused by the use of different Young's moduli for concrete in the calculations (normal value for temperature calculations and effective modulus for shrinkage calculations). When the steel section is slender, the compression due to shrinkage and temperature should be taken into account in checking the ultimate limit state. Equations 6.5 and 6.6 can be used for this purpose. For compact sections, the longitudinal stresses thus calculated are for checking the serviceability limit states only. In addition, the shear forces on the end connectors should be assessed by applying equation 6.9. Finally, the secondary effects on continuous beams can be calculated by applying the curvature from equation 6.11 to the simply supported beam and carrying out the appropriate indeterminate analysis for the support reactions. This enables the additional stresses and deflexions due to secondary effects to be determined.

The magnitudes of shrinkage strains are specified in codes of practice. The bridge Code lists values of between 100 and 300 microstrains for a given range of concrete mixes, guidance for concrete outside the given range being available from Appendix C of Part 4 of the Code. The distribution of temperature for design calculation is more complicated and is discussed in the following section.

6.1.4 *Determination of temperature on bridges*
Modern bridges tend to be continuous over longer spans. Thermal effects on these bridges could lead to the loss of serviceability and, in some cases, could and did cause failures of falsework during construction. Thermal behaviour of bridges has been given more attention in recent years, and adequate information has been incorporated into the Bridge Code, due mainly to the

work of Black and Emerson at the Transport and Road Research Laboratory.[26]

The distribution of temperature through a vertical section of a bridge deck is quite complicated and deserves some physical interpretation.[26] For this purpose, the distribution for a concrete deck 1 metre thick is shown in figure 6.5. The temperature within the top (about 0.5 m) of the deck is controlled by the incident solar radiation and drops as the distance from the sunlight increases. From this depth to about 0.3 m above the soffit, the temperatures are mainly dependent on those of the two previous days. Within the bottom 0.3 m, the temperatures depend on the shade temperature at the time and the amount of heat reflected from the ground beneath the bridge.

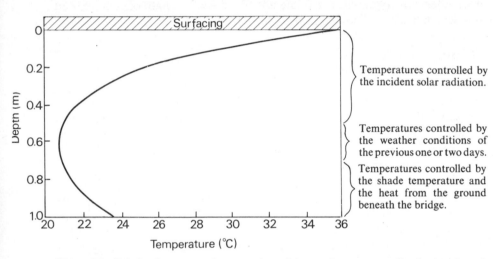

Figure 6.5 Relation between environmental conditions and temperature distribution through vertical section of bridge deck

For a composite deck, the lower part of the distribution is substantially different. Because the steel areas are shaded from the sun by the concrete slab, the temperature below 0.5 m from the slab soffit is controlled by the shade temperature. Between May and early August the shade temperature can vary over almost the full annual range. Figure 6.6(a) shows a temperature difference distribution (reducing temperature at the top to zero by adding a constant temperature) measured through the full depth of the deck of a composite bridge, the Adur Bridge main viaduct.

Another critical temperature distribution results from cooling of the deck surface, and is known as *reversed temperature distribution* (reversed being used to indicate a lower temperature at the deck surface). A typical distribution is shown in figure 6.6(b) for the Adur Bridge main viaduct.

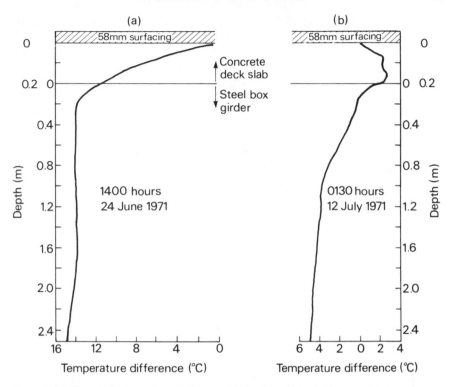

Figure 6.6   Measured temperature difference distributions (Adur Bridge—main viaduct)

The two critical distributions given by the Bridge Code (see figure 6.7) approximate closely to the above two measured distributions. In the Bridge Code, each temperature distribution is defined by two parameters—the thickness of the surfacing and that of the slab. The temperatures related to slab thicknesses can be obtained from figure 6.7, while the temperature corrections for surfacing thicknesses can be obtained from table 12 of the Code.

For the changes in the overall length of the bridge deck, as are required for the design of bearings and expansion joints, a single temperature known as the *effective bridge temperature* is used. The Bridge Code gives tables (10 and 11 in Part 2) for the minimum and maximum effective temperatures, which are related to shade air temperatures. It is therefore necessary first to find the minimum and maximum shade air temperatures from the maps of isotherms for the particular location of the bridge. These are then adjusted for the correct sea-level and the intended return period before obtaining the corresponding effective temperatures.

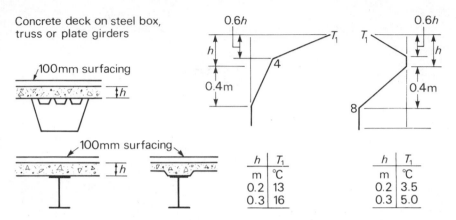

Figure 6.7   Temperature difference (BS 5400). Redrawn from BS 5400: "Steel, concrete and composite bridges, Part 2: Classification for loads", British Standards Institution 1978

6.2   Propped and unpropped construction

Previous discussion on stress calculations is based on the assumption that the slab and the steel beam act compositely at all times. During construction, however, the steel beam may carry the weight of the wet concrete slab, the formwork, plant and personnel, unless appropriate supports are provided under the steel beam to carry these loads. Such a provision of support is known as *propping* and is common practice in building. The number of such temporary supports need not be high. In fact, a single support may be acceptable if the stresses due to the construction loading are sufficiently low.

Another obvious form of "propped" construction is the casting of slab while the steel beam is supported by the ground. The composite beam is erected into position after the concrete has attained sufficient strength. This method is more suitable for bridge construction for which propping is invariably a costly operation.

In the majority of bridges unpropped construction seems to be the only practical solution. The steel section should then be designed to carry the appropriate load before the slab has hardened sufficiently. Under these circumstances, the top flange of the steel beam is subjected to relatively high compression, and care should be taken to brace the compression flange against buckling. To calculate the total steel stresses on subsequent loading, the stresses due to the construction loads on the steel beam have to be added to those due to the subsequent loads. On the other hand, the slab is free from stresses under the weight of the composite beam alone.

As far as stresses are concerned, unpropped construction amounts to a transfer of stresses from the slab to the steel beam and leads to a lower interface shear for the same superimposed load. The ultimate moment of resistance, however, is not affected provided that the section is compact. Otherwise (i.e. if it is a slender section) the higher steel stresses tend to cause premature local buckling of the steel beam.

6.3 Prestressing of composite beams

6.3.1 *Introduction*
The main objective of prestressing is to reduce or, in some circumstances, prevent the cracking of concrete under dead and imposed loads. The introduction of compressive stresses also enables the otherwise cracked concrete to develop effective stiffness, thus making the beam stiffer than its unprestressed counterpart. It is generally accepted that excessive cracking of the concrete increases the risk of reinforcement corrosion, but the relation between the actual extent of corrosion and the crack width has yet to be quantified.

Extensive research on the prediction of crack width in reinforced concrete members has been carried out in Britain, and the relevant information incorporated into current Codes of Practice. In general, British and American engineers are more tolerant towards cracking than Continental engineers. For the majority of bridges in Britain, limited tensile cracking of concrete decks is permitted, provided that adequate waterproof membranes are used and that the risk of membrane rupture due to concrete cracking is remote. In Continental Europe, however, the prevention of cracking by prestressing is the preferred solution, and current knowledge in prestressing is to a great extent based on Continental experiences.

6.3.2 *Method of prestressing*
Methods of prestressing vary from country to country and are therefore impossible to generalize. The current Bridge Code lists five methods for composite beams. The two underlying principles are the use of tendons to provide compression in the concrete, and the manipulation of reactions at supports on the beam to induce the desired compression. The five methods are:

(1) CASTING CONCRETE ON STEEL FLANGE IN TENSION
A moment is applied to the steel section to induce stresses with the same sign as those due to the design loading. The tension flange is then encased in concrete, and the moment released when the concrete has adequate strength.

The concrete is thus compressed in the absence of the imposed load and, in theory, will return to the stress-free state on application of the original moment.

(2) JACKING TO ALTER LEVEL OF SUPPORT
After part or the whole of the concrete deck has been cast and matured, the relative levels of the supports of a continuous beam can be altered by jacking. The required settlements to be imposed are calculated to induce the desired compression.

(3) INDEPENDENT PRESTRESSING OF SLAB
Prestressing the concrete slab or sections of the slab by tendons, or jacking whilst it is independent of the steel section and subsequently connecting them.

(4) INDEPENDENT PRESTRESSING OF STEEL BEAM
Prestressing the steel beam by tendons prior to concreting. The tendons may or may not be released after the concrete has matured.

(5) PRESTRESSING OF COMPOSITE SECTION
Prestressing the composite sections by tendons or jacking.

A well-known example for method (1) is the Preflex beam, first conceived by the Belgian engineer Lipski in 1949. The sequence of stresses is illustrated in figure 6.8 and described below:

Stage I: A steel beam is pre-loaded, usually by two loads at quarter span. The tension flange is then encased in high-strength concrete.

Stage II: After the concrete has gained sufficient strength, the loads are removed. The steel beam and the bottom slab act compositely, thus reducing the previous steel stresses and introducing compression into the bottom slab.

Stage III: When the Preflex beam is in position in the bridge, the top slab together with the encasement (to the web) is cast *in situ*. The dead load thus acts on the steel beam composite with the bottom slab.

Stage IV: For the superimposed dead load and live load, the member will act as the fully composite beam.

Stage I	Stage II	Stage III	Stage IV
Moment Pa on steel beam	Moment Pa on composite beam	Moment due to top slab and encasement	Moment due to live load

Figure 6.8   Various stages of stressing a Preflex beam

The main advantages of Preflex beams are increased stiffness under the range of working load, and the provision of encasement to the steelwork, with fewer cracks when compared with an encased beam of the same size.

Furthermore, since the steel beam has to be subjected to a high level of preflexion stresses during manufacture (Stage I), the material strength has virtually been proof-tested.

Another example of introducing precompression in the concrete slab is prestressing by cambering. This consists of setting up the steelwork above its intended level, prior to casting the concrete slab. After composite action is acquired, the bridge deck is lowered to the original level. The level required to be raised may become excessive in continuous multi-span bridges. To overcome this practical difficulty, cambering can be applied to sections of the bridge and hinges introduced between the section, as suggested by Roik.[27]

Method (2) is relatively simple in concept. Jacking up supports after the slab is cast is equivalent to the application of external reactions. Methods (3) to (5) refer mainly to the use of tendons to introduce compression in the slab. The relevant methods of stress calculation can be found in textbooks on the design of prestressed concrete structures.

6.4 Fatigue

6.4.1 Introduction
Consideration of fatigue is usually unnecessary in the design of buildings. In bridge design, however, because of traffic loading, fatigue becomes a major consideration. Fatigue damage is caused by a propagation of cracks under repeated loading, and welds are most vulnerable in this respect. Welding introduces stress concentrations, residual tensile stresses and defects which act as crack initiators. Figure 6.9 (BS 5400: Part 10: 1980) shows some welded details on which Codes of Practice require that fatigue assessment has to be made. Such requirements apply to steel deck and reinforcement as well, but those related to shear connectors are of particular relevance to composite construction. General design methods are given below to illustrate the principle of fatigue design, greater detail being available from the relevant Codes of Practice.

6.4.2 Stresses due to repeated loading
Because fatigue strength depends on the history of the applied stresses, much more information on loading is required than in static calculations. For a meaningful assessment of fatigue performance, the following information is needed and is given in the Bridge Code.

(a) Standard load spectrum. The weights of a number of specified commercial vehicles, together with their relative frequencies of occurrence.

(b) Annual flow of vehicles. Number of commercial vehicles assumed to travel along each lane per year. This is related to the category of road and whether it is a slow or adjacent lane (BS 5400: Part 10, Table 1).

(c) Standard fatigue vehicle with specified axle arrangement and weights. The purpose is to use one single vehicle for the group to simplify calculation.

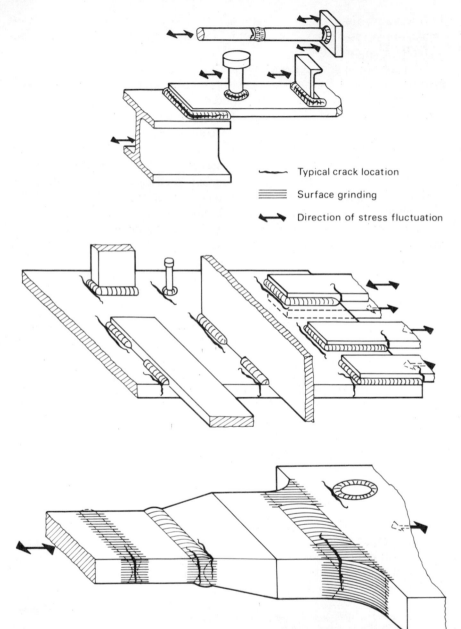

Figure 6.9   Welded details for fatigue assessment. Redrawn from BS 5400: "Steel, concrete and composite bridges, Part 10: Code of Practice for Fatigue", British Standards Institution 1980

The general "fatigue loading" at a point is known as a stress spectrum and consists of a finite number of stress ranges and the corresponding frequencies of occurrence. If the member is subjected only to tension and compression in one direction, the stress range at any point is the algebraic difference between the maximum stress and the minimum stress (e.g. adding the values if they are of opposite signs). Thus, by passing a specified commercial vehicle along the bridge, the influence line for the direct stress at the point can be obtained, and the stress range calculated from the peak and trough of the influence line. This stress range will have a frequency appropriate to the traffic of the vehicle over the bridge during the latter's designed life (120 years). The frequency can be calculated from the relative frequency for the particular vehicle in the group and the annual flow of commercial vehicles for the category of road in question (BS 5400: Part 10, Tables 11 and 1 respectively).

Repeating the above calculations of stress range and the associated frequency for other vehicles in the group, a design spectrum similar to that shown in figure 6.10 (eleven vehicles) will be obtained. Some complications

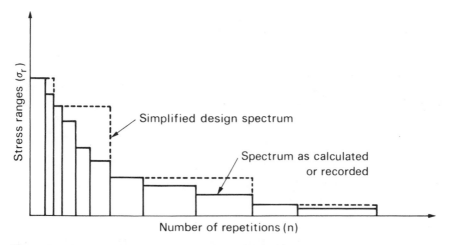

Figure 6.10 Simplification of a design spectrum. Redrawn from BS 5400: "Steel, concrete and composite bridges, Part 10: Code of Practice for Fatigue", British Standards Institution 1980

in the stress range calculation may arise when the directions of principal stresses vary with the passage of vehicles along the bridge (e.g. stress in parent metal).

However, the stress range can still be taken as the algebraic difference between the peak and trough stresses (from the influence line of principal stresses) provided that the principal directions of these two stresses are not more than 45° apart.

In assessing the fatigue life of stud connectors, it should be noted that crack initiation may occur in the parent metal or in the weld throat, as illustrated by figure 6.9. In the latter case, however, the stress range calculation is relatively simpler, since principal directions will not be involved, the stress in the weld being related to the longitudinal shear load on the stud (BS 5400: Part 10, Section 6.4.2).

6.4.3 Fatigue calculation

Vehicle spectrum method

Given the design spectrum (figure 6.10), this method is used to check whether fatigue failure will occur. First, assuming only one stress range is effective, the relevant number of repetitions to cause complete failure is calculated. This relation can be obtained from the well-known σ_r–N curve or from the following equation:

$$N\sigma_r^m = K \tag{6.21}$$

where K and m can be obtained from Codes of Practice for the given classes of welded detail (BS 5400: Part 10, Table 8).

If N_1 is the number of repetitions to failure for the particular stress range, then it follows that its frequency (say n_1) should not exceed N_1. In order to take account of the full spectrum, i.e. the cumulative effect of other stress ranges, Miner's summation is used, the safe condition being that the sum should be less than unity:

$$\sum \frac{n}{N} = \frac{n_1}{N_1} + \frac{n_2}{N_2} + \cdots + \frac{n_s}{N_s} < 1 \tag{6.22}$$

where s = number of stress ranges.

Simplified procedures

These methods are more conservative and depend on the availability of prescribed values of limiting stress range. The Bridge Code gives such limiting values for various categories of roads and welded details. The only major calculation required will be the application of the Standard Fatigue Vehicle to the slow and adjacent lanes to obtain the stress range. If the stress range is below the recommended limit, the fatigue life will be adequate.

6.5   Slender cross sections

The applicability of the simple plastic theory to composite beams depends upon one important assumption in respect of the steel beam—that substantial plasticity can develop without buckling. Substantial plasticity can be defined in terms of two requirements: sufficient yielding to resist a moment close to the fully plastic value, and sufficient deformation to sustain a hinge rotation. An ideal solution, as far as design is concerned, is to use simple calculations to identify the family of composite sections to which the simple plastic theory can be applied. These are known as *compact cross sections* while the others are *slender cross sections*.

When a composite section is subjected to negative bending moment over a continuous support, the flange of the steel beam is vulnerable to instability. For positive bending, the top flange and the web of the steel beam may buckle only if the neutral axis is in the web and vertical shear is significant. The classification is therefore related primarily to the loading condition. With the compressed portion of the steel section identified, a rigorous approach would consist of a stability analysis taking account of the appropriate boundary condition (i.e. with a free edge or restrained by a slab). To simplify this stability analysis, the Bridge Code recommends the use of length/thickness ratios of the compressed panels for the stability assessment. The same criteria as those for steel beams are used so that Part 3 of BS 5400 is applicable. The results tend to be over-conservative, mainly because the increase of stability due to the presence of the slab is ignored.

Although the Bridge Code allows the use of fully plastic moments in simply supported beams when the neutral axis lies within the slab or the compression flange, the reduction in the fully plastic moment due to vertical shear has to be taken into account. A rigorous method for this reduction consists of combining the vertical shear stress and the longitudinal stress to modify the distribution of constant yield stress in accordance with the yield criterion. Alternatively, the effective area for resisting vertical shear can be taken as the web area, less that portion adjacent to the top flange where the longitudinal tensile strain exceeds the yield strain.

For slender cross sections, the calculation of the ultimate moment is based on an elastic section modulus approach, with fibre stresses of steel and concrete given by the Codes of Practice. Accordingly, the slab should also be designed elastically and not (for example) based on the yield-line theory. Furthermore, the amount of moment redistribution at the ultimate limit state is lower for the slender continuous beam, the allowable percentages being given in Codes of Practice (also see section 4.2.1).

6.6 Control of cracking in concrete

Excessive cracking of concrete adversely affects the appearance and durability of the structure, and provisions should be made in design and construction to minimize the extent of cracking. This problem is unlikely to arise in the design of buildings unless the top surface of the composite T-beam is exposed to corrosion. In the design of bridges, which are vulnerable to corrosion by de-icing salt, the designer has to consider the requirement of the particular bridge and its environment, so as to arrive at a specific limiting crack width. On the other hand, since cracks open and close continually with the passage of traffic, and it is the long-term exposure which is believed to do the damage, the designer can ignore abnormal loads (HB loading) in the relevant calculation.

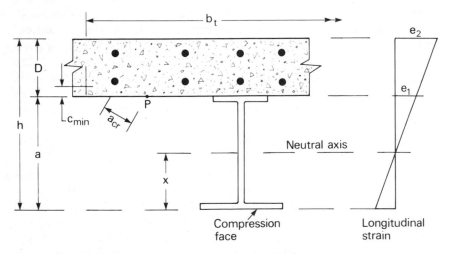

Figure 6.11   Crack-width calculation for a point on the slab of a composite section

An easy way of satisfying the serviceability criterion in relation to cracking is to adopt bar spacings recommended by Codes of Practice. Bar-spacing rules, however, are inevitably conservative. It may also happen that a preliminary calculation is sufficient to show that a given design is satisfactory without the full calculation of the crack width (i.e. zero crack width). Referring to figure 6.11, which shows a composite section under a hogging moment together with the strain distribution based on the "cracked" section, the criterion of "zero crack width" is

$$e_3 = e_2 - 0.0012 \, A_c/F_s \leqslant 0 \qquad (6.23)$$

where A_c = area of slab (in mm^2)
           = effective breadth × slab thickness
     F_s = (area of reinforcement + area of any encased tension flange) × characteristic yield stress of reinforcement—in N/mm^2
     e_3 = fictitious strain to assist calculation.

Crack-width formula

The approach is taken from Part 5 of the Bridge Code but some rearrangements have been made to simplify calculation. It is based on the assumption that the crack width at the point on the member subjected to bending depends on three factors:

   (i) The distance (a_{cr} in figure 6.11) from the point at which the crack width is required to the nearest bar placed perpendicular to the plane of the crack (i.e. the bar provided to resist bending moment in the same direction as the cracking moment).

  (ii) The distance from the point to the neutral axis, i.e. ($a'-x$).

 (iii) The average surface strain at the point when cracking is considered as a random process. The implication is that a crack may or may not develop at the point. For samples in which the point lies between two cracks, the strain based on the cracked section (e_1) will be overestimated. Hence the stiffening effects should be allowed for by reducing e_1. The average strain (e_m) is given by equation 6.25.

The crack width at a point is given by the following formula, some symbols having already been defined (figure 6.11 in which cracking at the bottom of the slab is checked):

$$w = \frac{3a_{cr}e_m}{1+2(a_{cr}-C_{min})/(h-x)} \tag{6.24}$$

$$e_m = \frac{a'-x}{h-x}\,e_3 \tag{6.25}$$

where C_{min} = minimum cover to the reinforcement,

        x = depth of neutral axis from the extreme compression fibre (compression face),

        a' = distance from the point to the compression face.

Since the neutral axis rarely lies within the slab for negative bending on the cracked section, the fraction in equation 6.25 is positive, and e_m and e_3 have the same sign. When the calculation shows that the tensile strain e_m is negative, it implies that there will be no cracking, hence the simplified criterion of equation 6.23.

A further criterion is that the stress in the tensile reinforcement should not be excessive, and is governed by the serviceability requirements of the relevant Codes of Practice (for example, Part 4 of the Bridge Code).

7   Composite Columns

7.1 Introduction

By definition a composite steel-concrete column is a member with a cross-section consisting of a steel section (or sections) and concrete which act together to resist axial compression. The choice, however, is restricted in practice by current experience in design and construction. Only two common types of composite column are in use—steel sections encased in concrete and hollow sections filled with concrete (figure 7.1). Until the 1950s, steel stanchions were usually encased in low-strength concrete primarily to provide fire protection. Modern design of composite columns makes use of both the

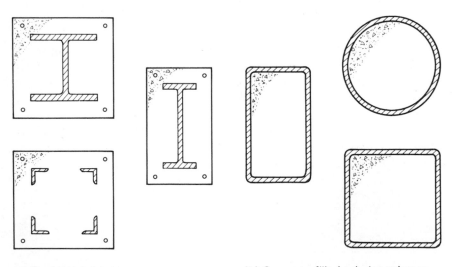

(a) Encased columns   (b) Concrete-filled tubular columns

Figure 7.1   Types of composite column

strength and the fire resistance of concrete and therefore provides a more economical structure.

Columns are subjected mainly to axial loads and, compared with beams, the transverse shear (change in bending moment along the length) is much lower. It is therefore not surprising to find that mechanical shear connectors are normally unnecessary to develop complete interaction in composite columns. Experiments on encased sections have shown that the interface bond depends on the thickness of the concrete cover and the amount of reinforcement. Additional relevant factors such as surface finish and ratio of diameter to wall thickness have been identified in further tests on concrete-filled tubes. When concrete cover and cleanliness at the interface comply with the relevant codes of practice, the interface bond can be assumed satisfactory without the provision of any mechanical connectors. However, in special circumstances (such as the occurrence of significant interface shear) a calculation will be appropriate to see if the limiting stress given in the code is likely to be exceeded.

7.2 Methods of design

Because of inelasticity of materials and the increase of external moment with deflexion, the mathematical solution to the column problem and the method of design (method of analysis, to be exact) are relatively complicated. The simplest method in current codes of practice is the Case-Strut Method (BS 449: 1959) which is very conservative. This method makes use of a design rule for uncased steel struts and introduces a modification of a parameter to take account of the presence of concrete.

For a steel strut subjected to axial load (N) and bending leading to compressive stress (f_f) (including bending about both axes), the following condition must be observed:

If only axial load is acting,

$$\frac{N}{N_a} \leqslant 1 \tag{7.1}$$

where $N_a = A_s f_{ac}$
f_{ac} = allowable axial stress.

If only bending moments are acting,

$$\frac{f_f}{f_{af}} \leqslant 1 \tag{7.2}$$

where f_{af} = allowable compressive stress (for bending).

When both axial load and bending are acting on the steel strut, the following condition will apply:

$$\frac{N}{N_a} + \frac{f_f}{f_{af}} \leqslant 1 \tag{7.3}$$

The presence of concrete is allowed for in two ways. The allowable axial load (N_a) is increased by considering the additional strength provided by the concrete encasement. The concrete encasement also increases the radius of gyration, which in turn will result in a higher allowable axial stress. Further details can be obtained from BS 449: Part 2. Tests on cased struts have shown that this method is grossly over-conservative and inaccurate, mainly because it is hardly possible for such a simple formula to take account of the large number of relevant parameters.

Research on improved methods of design of composite columns has been undertaken in the United Kingdom since the early 1960s, mainly at Imperial College, London. Following the early series of tests by Stevens[28], [29] at the Building Research Station and by Bondale[30] at Imperial College, an extensive theoretical study on design methods was conducted by Basu. In 1969, Basu and Sommerville[31] published a method based on the Perry-Robertson type of buckling curve, applicable to pin-ended composite columns, which has since became the basis of current methods of design. In the mid-1970s, while a new generation of structural codes was being conceived in the United Kingdom and harmonization of European Codes[32] was under way, the need to co-ordinate national and international codes became more important than ever. It is in respect of this need that the method proposed by Virdi and Dowling[33] should be regarded as a significant milestone. Apart from the improvement in accuracy, their method has bridged the gap between composite column design and steel column design, in that a set of steel column curves was adopted as a common basis for both

Table 7.1 Classes of composite columns

Classification of columns for analysis			Section
pin-ended	axially loaded	short columns	7.3.1
		slender columns	7.3.2
	with end moments	uniaxial bending	7.4.1
		biaxial bending	7.4.2
with end restraints			7.5
with lateral sway			7.5
with transverse loads			7.5.1

types of column. The same approach was adopted by the European Convention for Constructional Steelwork (ECCS) to unify the basis of design for composite and steel columns.

To assist understanding of this somewhat complicated subject, the various methods will be progressively developed in the order of increasing complexity. Table 7.1 shows the various methods applicable to the appropriate classes of column. The ultimate limit state appears to be the only major criterion of design. No specific criteria for the serviceability limit state are given in current codes of practice in relation to composite column design.

7.3 Pin-ended columns with axial load only

7.3.1 *Squash loads*
Composite columns referred to hereinafter are of uniform cross-section symmetrical about two perpendicular axes. Unless otherwise stated, they are designed to resist short-term static loads, and the effect of loss of interaction on the ultimate capacity can be neglected. Encased columns and concrete-filled rectangular columns will be dealt with here, while concrete-filled circular columns will be discussed in section 7.6.

In the study of structural stability, the term "short column" refers to a compression member which can attain its ultimate capacity, known as the *squash load*, without buckling. By implication, the column must be perfectly straight and subjected only to axial load. Since the strains in steel and concrete at any cross-section are equal, the steel will yield first as the load increases. Because of the ductility of steel, this steel force (due to the steel section and the reinforcement) can be maintained as concrete is developing its maximum strength under increasing load. Assuming no local buckling or torsional buckling occurs, the maximum stresses of the steel and concrete will develop. Hence the ultimate load (squash load) to be used in design can be related to the design strengths of the concrete, steel and reinforcement by the following equation (the symbol f_y is not used because there are two different values for the steel section and the reinforcement, hence the suffixes s and r):

$$N_u = A_c f_c + A_s f_s + A_r f_r \qquad (7.4)$$

A_c = cross-sectional area of concrete, which may be taken as the gross area since the areas displaced by the steel section and reinforcement are comparatively small,

A_s = cross-sectional area of the steel section,

A_r = total cross-sectional area of the reinforcement,

f_c = design strength of concrete in direct compression, which is obtained by dividing the cube (or cylinder) strength by the appropriate factor given by the relevant code,

f_s = design yield strength of the structural steel, again obtained from the yield strength and a given factor,

f_r = design yield strength of reinforcement, again as a factored yield strength,

It has been pointed out that the squash load can only be attained under ideal conditions. The column must be perfectly straight, and the absence of eccentricity of loading and of bending moments has to be assumed. After the steel section and the reinforcement have yielded and the concrete is being loaded into the region of reducing stiffness, the overall reduced stiffness is assumed adequate to prevent local and torsional buckling.

Tests have shown that composite columns capable of reaching their squash loads are too short to have any practical significance. In fact, the term *short column* has now acquired a different but incorrect meaning for design purposes, as in the composite Bridge Code BS 5400: Part 5: 1979. Although not explicitly defined in the Code, the short column is assumed to be an axially loaded member which fails by buckling about the minor axis due to initial imperfections in the straightness of the steel member; any unintended moments due to construction tolerances can be adequately allowed for by reducing the buckling load by a fixed percentage. For columns beyond the defined range (slender columns), the unintended moments due to construction tolerances have to be treated as an eccentricity which increases bending about the minor axis as the load increases. This is a realistic way of allowing for the increasing effect of tolerances as columns become more slender.

The buckling load of short columns will now be discussed. In order to minimize the confusion due to the two definitions, loads associated with these columns will be used. Squash loads will be related to the ideal short column, and buckling loads will include those of short columns as used in codes of practice. With this clarification, references to the ideal short column will no longer be necessary.

Before leaving this section, it is appropriate to introduce a factor which indicates the proportion of squash load carried by the concrete alone. It is known as the *concrete contribution factor* and is given by

$$\alpha = A_c f_c / N_u \qquad (7.5)$$

This factor will be referred to in subsequent sections. It is hardly necessary to place limits on this factor, since there is no evidence of any significant change in behaviour due to its variation. However, some codes do give limits mainly to show the validity of the recommended methods of design.

7.3.2 Buckling loads

For axially loaded columns with pinned ends, it has been shown that the ultimate load can be obtained by simple calculation when they are extremely short. Taking the other extreme, ultimate loads of very slender columns can be calculated by assuming that buckling occurs (about the minor axis) while the columns remain elastic. A formula for this elastic buckling load, derived from elastic stability analysis, and known as the Euler load is given by

$$N_E = \frac{\pi^2(EI)}{l^2} \tag{7.6}$$

where (EI) = elastic bending stiffness, i.e. moment divided by curvature,
   l = length of column.

Alternatively the Euler load can be expressed in terms of the slenderness ratio l/r:

$$N_E = \frac{\pi^2(EA)}{(l/r)^2} \tag{7.7}$$

(EA) = axial load divided by the corresponding axial strain, or reciprocal of axial strain per unit load,
   r = radius of gyration (about minor axis),
   $(r^2$ = moment of inertia divided by area of section.)

For a given steel or composite section, the above two parameters, the squash load and the Euler load, can therefore be used to obtain the first approximation for the buckling curve (see figure 7.2). The figure shows a relation between the maximum load (N_m) and the slenderness ratio (l/r). The portion BC is the Euler curve, while the horizontal line AB implies that the maximum load is equal to the squash load (N_u in equation 7.4). For a steel column, curve BC becomes a simple equation independent of the shape of the section:

$$K_1 = \frac{N_m}{N_u} = \frac{\pi^2 E_s A_s}{(l/r)^2} \times \frac{1}{A_s f_s} = \frac{\pi^2 E_s}{f_s (l/r)^2} \tag{7.8}$$

Because of initial imperfections and material yielding, the axial maximum loads are lower than the Euler values, hence the Perry-Robertson curve shown by dotted lines in the figure has been used for steel struts.

The assumption that column strength depends only on slenderness ratio and squash load, as implied in the Perry-Robertson approach, is not strictly correct but provides a convenient method of design. Adopting the same approach, Basu and Sommerville carried out a theoretical study on a large number of axially loaded composite columns and obtained a curve lying below the Perry-Robertson curve (figure 7.2). Although the curve is

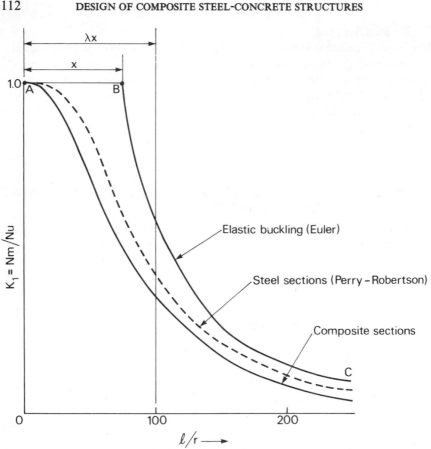

Figure 7.2   Column buckling curves

satisfactory for all practical purposes, it should be noted that, for large values of slenderness ratio, it implies that an encased section may have a lower capacity than the corresponding uncased section.

The method proposed by Virdi and Dowling is an improvement in accuracy and in the ease of application. Firstly, a new parameter, known as the *slenderness parameter*, is suggested to characterize a column (λ in figure 7.2, defined as the slendernesss ratio divided by AB). The slenderness ratio is in fact not quite ideal in characterizing a column; for example, it is impossible without further calculation to say if the slenderness ratio is on the left or right side of B. The use of the slenderness parameter gives an obvious indication as to how slender the column is relative to AB. For a given composite section, a column length can be found at which its elastic buckling load is equal to the squash load ($\lambda=1$). It is useful to note that composite columns with $\lambda=1$ fail at about half their full strengths (squash loads), while short columns with $\lambda=0.5$ and slender columns with $\lambda=1.5$ fail at about 90% and 30% of their full strengths respectively.

It is also interesting to note that the relation between K_1 and λ is independent of the properties of the composite section, whereas that between K_1 and the slenderness ratio (equation 7.8) is related to Young's modulus and yield stress. To find the $K_1 - \lambda$ relationship, the value of x (length AB) is first obtained by finding the intersection between AB and the Euler curve (i.e. setting K_1 in equation 7.8 equal to unity):

$$\pi^2 E_s = f_s x^2 \tag{7.9}$$

Equation 7.8 can now be expressed in terms of λ as

$$K_1 = \frac{\pi^2 E_s}{f_s(\lambda x)^2} \tag{7.10}$$

Eliminating x from the above two equations, the invariant relation between K_1 and λ is obtained:

$$K_1 = \frac{1}{\lambda^2} \tag{7.11}$$

To find the column length at which the Euler load equals the squash load, equations 7.4 and 7.6 are used:

$$N_E = \frac{\pi^2 (EI)}{l_E^2} = N_u \tag{7.12}$$

$$l_E = \pi \sqrt{\frac{(EI)}{N_u}} = \pi \sqrt{\frac{E_c I_c + E_s I_s + E_r I_r}{N_u}} \tag{7.13}$$

where l_E = column length at which the Euler load and squash load are equal,

I_c, I_s, I_r are moments of inertia about the axis of symmetry for concrete, structural steel and reinforcement respectively.

For a given radius of gyration (i.e. a given cross-section), the slenderness parameter is also a ratio of column length:

$$\lambda = \frac{l}{l_E} \tag{7.14}$$

The significance of adopting the slenderness parameter lies in the existence of $K_1 - \lambda$ curves for steel struts. Experimental and theoretical work on axially loaded struts has shown that their buckling loads can be satisfactorily predicted from the corresponding values of λ. The results have also established that, for better accuracy, three curves should be used (instead of one) to reflect the differences in geometry of cross-section, magnitude and distribution of residual stresses and initial imperfections in straightness. The use of three curves for three groups of steel struts has been found acceptable in the United Kingdom and Europe (ECCS). These curves, identified as a, b and c (figure 7.3) can also be approximately represented by an equation. They have been found to apply equally well to composite columns.

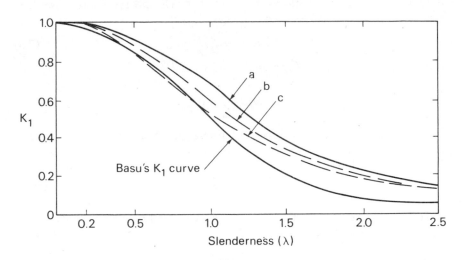

Figure 7.3 European curves a, b, c and Basu's K_1 curve

The design procedure for axially loaded composite columns can now be summarized as follows (see example 8.4.3):

(1) Calculate the slenderness parameter from equations 7.13, 7.14 and 7.4.
(2) From the geometry of the steel section and the direction of bending (axis xx or yy), use figure 7.4 to select the appropriate curve, or table as in the Bridge Code.
(3) Obtain from the correct table (or curve) the value of K_1 corresponding to the slenderness parameter.
(4) The buckling load is the product of K_1 and the squash load (equation 7.4).

Virdi and Dowling examined the validity of the above method by comparing their predicted values with failure loads of a wide range of axially loaded composite columns, obtained by theory and by test. The results showed excellent agreement with theory and realistic comparison with tests. The comparisons with theoretical results are shown in figures 7.5 to 7.7 under three separate groups (a, b, c). Similarly the comparisons with test results are shown in figures 7.8 to 7.10. The curves a, b and c in these figures can be used in the design procedure (see step 2 above).

Shape of steel section		Curve	Table

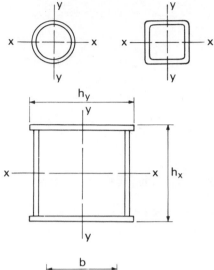

	Curve	Table
Rolled tubes Welded tubes (hot finished)	a	13.1

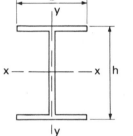

| Welded box sections
x–x, h_x
y–y, h_y | b | 13.2 |

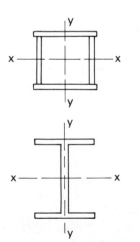

Rolled sections*		
Buckling about x–x axis		
$h/b > 1.2$	a	3.1
$h/b \leqslant 1.2$	b	13.2
Buckling about $y \sim y$ axis		
$h/b > 1.2$	b	13.2
$h/b \leqslant 1.2$	c	13.3
Welded sections		
Buckling about x–x axis		
(a) flame cut flanges	b	13.2
(b) rolled flanges		
Buckling about y–y axis		
(a) flame cut flanges	b	13.2
(b) rolled flanges	c	13.3

Box sections stress relieved by heat treatment	a	13.1

I and H sections, stress relieved by heat treatment		
Buckling about x–x axis	a	13.1
Buckling about y–y axis	b	13.2

*Not applicable to sections where the flange thickness exceeds 40 mm.

Figure 7.4 Strut curve selection chart. Redrawn and adapted from BS 5400: "Steel, concrete and composite bridges, Part 5: Code of Practice for Design of Composite Bridges", British Standards Institution 1979

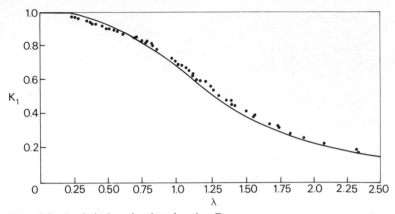

Figure 7.5   Analytical results plotted against European curve *a*

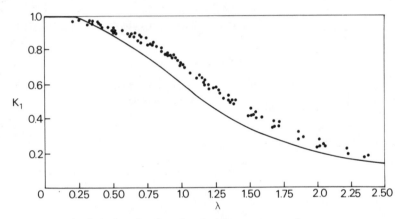

Figure 7.6   Analytical results plotted against European curve *b*

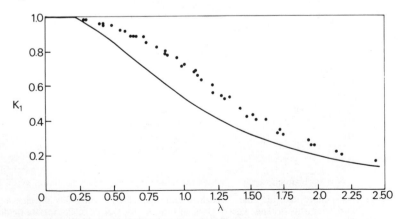

Figure 7.7   Analytical results plotted against European curve *c*

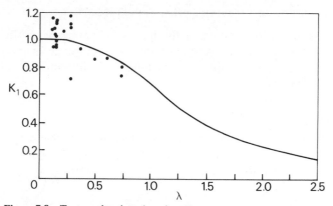

Figure 7.8   Test results plotted against European curve *a*

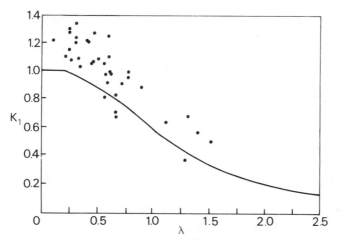

Figure 7.9   Test results plotted against European curve *b*

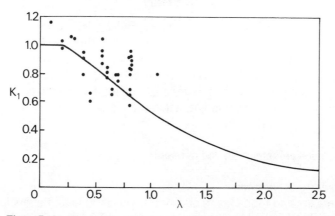

Figure 7.10   Test results plotted against European curve *c*

To discuss the determination of the numerical values of all the relevant parameters involved in the proposed method, reference is made to the summary of the design procedure and the following parameters are identified:

column length
area, moments of inertia
Young's moduli, design strengths

Only the design strength and Young's modulus for concrete deserve further comment, the meaning of the rest being unambiguous. The design strength of concrete in the column is related to the cube strength by

$$f_c = k_1 f_{cu}/\gamma_m \qquad (7.15)$$

where f_{cu} = cube strength,
γ_m = partial safety factor for concrete, which varies between 1.3 and 1.6 for the ultimate limit state but is specified in a given code,
k_1 = factor to take account of the discrepancy between the strength of concrete in the column and that in a cube; it varies between 0.67 and 0.70 (specified by code).

The evaluation of the Young's modulus appropriate to the concrete is less straightforward, since this is not a unique value but decreases as the strain gets larger. Virdi and Dowling suggest that since it is related to the Euler load, the use of the secant modulus would be more appropriate than the tangent modulus. In spite of this argument, they made some comparison on the basis of the initial moduli given by the following two equations:

$$E_{co} = 1000 f_c \qquad (7.16)$$

which is suggested in the CEB Model Code, and

$$E_{co} = 67\,200\sqrt{f_c} \qquad (7.17)$$

which is derived from CP 110 and where f_c and E_{co} are expressed in N/cm^2.

Having obtained satisfactory comparisons, they then used half the values of the Young's modulus in equation 7.16, which is equivalent to the secant modulus, and concluded that the design based on this reduced modulus would be on the safe side. The Bridge Code also recommends the use of a reduced modulus in connexion with such calculations of buckling load.

7.4   Pin-ended columns with end moments

7.4.1   *Bending in one plane*
When there are end moments in addition to the axial load, the buckling load as obtained by the method described in section 7.3 has to be reduced. The

method to be presented will use the previous buckling load as a starting point. Accordingly, the column shown in figure 7.11 differs from the previous model only in respect of the presence of end moments. Positive moments are as shown, and the moment at the lower end is the smaller of the two. For single curvature bending, β is positive.

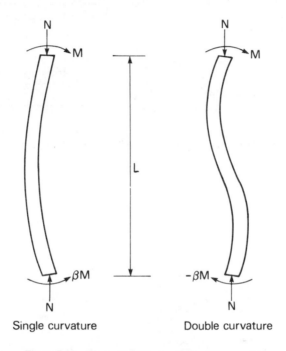

Figure 7.11   Pin-ended columns with end moments

The column behaviour can be described in terms of an interaction diagram showing the reduction in buckling load with increasing moment. One obvious limiting condition is pure bending under the fully plastic moment (M_u) and zero buckling load. A first approximation to the interaction curve can be obtained by ignoring buckling effects, and consists simply of an analysis of fully plastic sections. Taking an arbitrary plastic neutral axis, and assuming a fully plastic section, the moment and axial compression calculated from the stress blocks will give one point on the approximate (upper-bound) interaction curve.

The calculation is illustrated by figure 7.12 which shows stress blocks for the concrete in compression, the steel section and the reinforcement. The stress blocks A and B represent two equal and opposite forces, and are inserted to simplify calculation, since the forces in the steel section are then

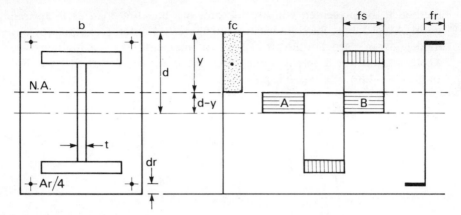

Figure 7.12   Section subjected to bending and compression (NA in web)

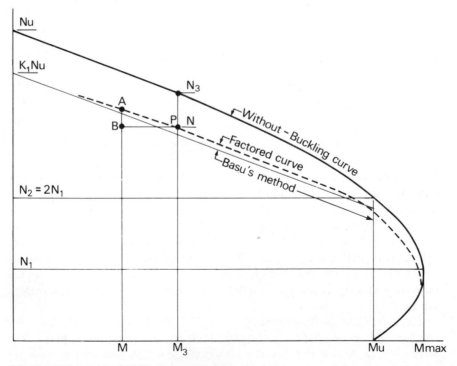

Figure 7.13   Axial load–end moment interaction curves

equivalent to a plastic symmetrical bending plus a single force twice the magnitude of A.

The resultant compression is

$$N = f_c by - 2f_s t(d-y) \tag{7.18}$$

The moment is the sum of those due to the concrete, the steel section and the reinforcement:

$$M = M_c + M_s + M_r \tag{7.19}$$

where

$$M_c = f_c by(d-y/2) \tag{7.20}$$

$$M_s = M_p - f_s t(d-y)^2 \tag{7.21}$$

$$M_r = f_r A_r(d-d_r) \tag{7.22}$$

M_p = plastic moment of the steel section alone.

By varying the position of the plastic neutral axis, sets of (M, N) values will be obtained to complete the interaction curve marked "without buckling" on figure 7.13. If the composite column is sufficiently short or carrying a very light axial load, then this curve can be used to check whether the co-existing load and moments would cause failure. The procedure consists of marking a point corresponding to N and M on the diagram, points lying outside the envelope indicating failure.

Taking the same composite section, but now specifying the column length and the proportion of the end moment (i.e. find l and β), sets of M and N can be obtained by a series of inelastic analyses, leading to the "actual curve" on figure 7.14. As expected, the solution of zero bending moment coincides with the value of K_1 discussed under section 7.3, and the solutions for small axial loads coincide with those ignoring buckling effects.

Basu and Sommerville studied about 100 such interaction curves (ABD) and suggested the use of a parabola passing through ABC (but not shown in the figure) and a vertical line CD (shown as dotted line in figure 7.14) to approximate the interaction curve. The parabola can be determined from three known points which in turn are fixed by the values of K_1, K_2 and K_3. They gave empirical formulae for K_2 and K_3 in terms of α (equation 7.5), β and the slenderness ratio of the composite column. The following formulae, however, make use of the slenderness parameter λ (equation 7.14) suggested by Virdi and Dowling:

$$\frac{K_2}{K_{20}} = \frac{90 - 25(2\beta - 1)(1.8 - \alpha) - C_4\lambda}{30(2.5 - \beta)} \tag{7.23}$$

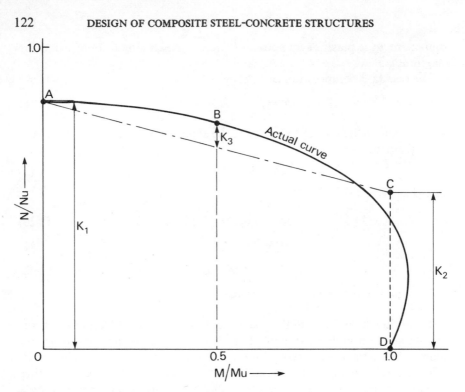

Figure 7.14   Idealized interaction curve for column design

$$K_{20} = 0.9\alpha^2 + 0.2 \tag{7.24}$$

$$0 \leqslant \frac{K_2}{K_{20}} \leqslant 1$$

(If K_2 is negative, it should be taken as zero.)

$$K_{20} \leqslant 0.75$$

where C_4 is taken as:

100 for columns designed on the basis of curve *a*
120 for columns designed on the basis of curve *b*
140 for columns designed on the basis of curve *c*

$$K_3 = 0.425 - 0.075\beta - 0.005C_4\lambda \tag{7.25}$$

and should lie between $-0.03\,(1+\beta)$ and $(0.2 - 0.25\alpha)$. For bending about the strong axis of the steel section, K_3 should be taken as zero.

Summary of checking procedure

(1) Work out the following parameters for the composite columns, using equations 7.4, 7.5, 7.13 and 7.14: N_u, α, λ.
(2) Find K_1 according to section 7.3.2.

(3) Calculate the ultimate moment for pure bending M_u, K_2 and K_3.
(4) Construct a parabola to pass through three points and complete the interaction curve (figure 7.14); the parabola is given by

$$\frac{N}{N_u} = K_1 - (K_1 - K_2 - 4K_3)\left(\frac{M}{M_u}\right) - 4K_3\left(\frac{M}{M_u}\right)^2 \tag{7.26}$$

(5) Check if the point M/M_u, N/N_u is inside the envelope of the interaction curve (M and N are the external moment and axial load, see figure 7.11).

Simplified methods

The method outlined above appears to be quite complicated for the purpose of checking that failure does not occur under the given axial load and moments. Two simplified methods will now be suggested. In working out the parameter for the interaction curve, no advantage has been taken of the fact that under small axial loads, buckling can be ignored. The maximum moment ($M_{\max}$), the ultimate moment (M_u) and the two axial loads (N_1 and N_2) corresponding to these moments (figure 7.13) can be obtained by analysis on the plastic section only. To calculate N_1 and $M_{\max}$, the neutral axis must be such as to make dM/dN zero (where the vertical tangent touches the interaction curve). From equations 7.18 to 7.22 (and figure 7.12),

$$\frac{dM}{dy} = (bf_c + 2tf_s)(d-y) \tag{7.27}$$

$$\frac{dN}{dy} = bf_c + 2tf_s \tag{7.28}$$

Hence, the neutral axis for maximum moment is given by

$$\frac{dM}{dN} = \frac{dM}{dy} \div \frac{dN}{dy} = d - y = 0 \tag{7.29}$$

The last equation establishes a useful rule: that the maximum moment is obtained when the neutral axis coincides with the plastic neutral axis of the steel section. From the same set of equations, simple expressions for the maximum moment and N_1 can be obtained:

$$M_{\max} = \tfrac{1}{2}f_c bd^2 + M_p + M_r \tag{7.30}$$

$$N_1 = f_c bd \tag{7.31}$$

By inspection of equations 7.27 and 7.28 (integrating with respect to y), it can be seen that the $N-M$ envelope is a parabola. Hence N_2 is twice N_1, and should be used to give the following approximate value for K_2 (figure 7.14):

$$K_2 = \frac{N_2}{N_u} = \frac{2f_c bd}{N_u} = \frac{A_c f_c}{N_u} = \alpha \tag{7.32}$$

This provides the basis for the first approximate method, in which K_3 is ignored so that the interaction curve is represented by two straight lines (AC and CD in figure 7.14). Hence, the failure-boundary line given below defines the combination of M and N which causes column failure:

$$\frac{N}{N_u} = K_1 - \frac{M}{M_u}(K_1 - \alpha) \tag{7.33}$$

It is quite satisfactory to ignore K_3 for the purpose of checking, since it would in most cases be on the safe side (unless K_3 is negative). In fact, K_3 is often zero and, even when it is negative, its magnitude never exceeds 6% of N_u.

An alternative is to use the "without buckling" curve as a basis. One assumption will be made—that for a given moment, the axial loads from the "actual curve" and the "without buckling" curve are at a constant ratio (K_1). It is found that this factored curve is very close to the actual curve and can be used as the failure envelope. To check if a set of co-existing load and moment represented by N and M (point B on figure 7.13) is a safe set, a horizontal line is drawn through B to meet the factored curve at P. If P is on the right side of B, the column strength is adequate. This criterion can also be explained by observing that the point P representing N, M_3 is on the failure envelope. Since point B represents a combination of the same axial load with a smaller moment (N, M), the column strength must be adequate.

The procedure consists of checking that M_3 exceeds M and no curve-plotting is involved. The value of M_3 can be obtained from the plastic-section analysis (figure 7.12) since (N_3, M_3) is on the without-buckling curve. Making use of the factored relation to calculate N_3, equation 7.18 can be applied to find the neutral axis and hence moment:

$$N_3 = N/K_1 \tag{7.34}$$

$$y = \frac{N_3 + d(2tf_s)}{bf_c + 2tf_s} \tag{7.35}$$

$$M_3 = \tfrac{1}{2}bf_c y(2d-y) - tf_s(d-y)^2 + M_p + M_r \tag{7.36}$$

If the calculated value of y indicates that the neutral axis is in the bottom flange, the following two equations should be used instead:

$$y = (C + 2f_c b_f(2d - d_s) - A_s f_s)/(f_c b + 2f_s b_f)$$

$$M_3 = f_c b y(d - y/2) + f_s b_f(y - d_s)(2d - y - d_s) + M_r$$

where b_f = breadth of steel flange
d_s = thickness of concrete cover to steel section.

7.4.2 *Biaxial bending*

This section deals with columns which can fail by buckling about either axis. A column subjected to axial load and bending about the major axis only could nevertheless buckle about the minor axis, because of the initial imperfections in the straightness of the steel member. Hence the previous interaction curve for the uniaxial bending about the major axis (OEGC in figure 7.15) would have to be modified. Curve OEC is replaced by OED, because as axial load increases above the point G, buckling failure about the other (minor) axis takes over. The interaction curve for axial loads and uniaxial bending about the minor axis, OFD, is not affected because without bending moment about the other axis, failure will occur about the minor axis only. Current methods of design make use of column strengths corresponding to these two distinct cases of uniaxial bending, and approximate formulae are given to calculate the biaxial column strength.

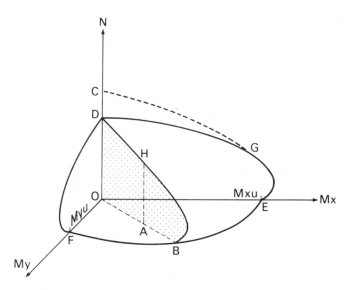

Figure 7.15 Biaxial buckling surface showing safety space for sets of M_x, M_y, N

It should be noted that column strength is used in place of the simpler checking procedure. Referring to figure 7.14, checking involves confirming that the point (M, N) is inside the envelope, whereas determination of the strength of the column consists of finding the magnitudes of the load and moment under which failure occurs. The failure condition depends on the way the existing load and moment are increased—proportionally or one being held constant. Proportional loading appears to be more realistic, but current codes allow the use of a constant moment, thus enabling a higher axial load to be used (point A on the actual curve, figure 7.13).

In their study on an interaction formula for biaxial bending, Virdi and Dowling tested a series of concrete encased H-section columns with varying length and biaxial eccentricities.[34] An analytical method was developed in parallel for the computation of the failure load, taking account of residual stresses in the steel section and of the lack of initial straightness. Having obtained a satisfactory agreement between the experimental and theoretical results, further computing was carried out to cover a practical range of columns. The following equation for column strength under biaxial bending was proposed:

$$\frac{1}{N_{xy}} = \frac{1}{N_x} + \frac{1}{N_y} - \frac{1}{N_{ax}} \tag{7.37}$$

where N_{xy} = failure load under biaxial bending,

N_x = failure load under uniaxial bending about the major $(x–x)$ axis,

N_y = failure load under uniaxial bending about the minor $(y–y)$ axis,

N_{ax} = failure load under axial load only, assuming buckling about the minor axis to be prevented.

This formula has been shown by test and theory to be safe for design application. In fact, for slender columns, results have been shown to be unduly conservative (design strength being only half of test and computer values). The equation has been incorporated into the Bridge Code BS 5400: Part 5: 1979.

Simplified method

The basic principle of the simplified method for biaxial bending is given below. If desired, some variation can be easily incorporated to improve its accuracy. The effects of the smaller end moments are ignored, but the appropriate formulae containing β can be used to take account of these effects. The problem is to check if a given column fails when subjected to an axial load and two moments with values N, M_x and M_y. A simple start is to see if the biaxial moments alone would cause failure. No stability is involved here, and the following approximate equation of an ellipse (figure 7.16) can be used for the failure envelope (exact for a circular or tubular section): for any point B on the curve,

$$\left(\frac{M_{xB}}{M_{xu}}\right)^2 + \left(\frac{M_{yB}}{M_{yu}}\right)^2 = 1 \tag{7.38}$$

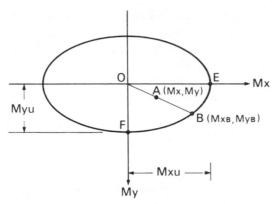

Figure 7.16 Elliptical failure boundary for biaxial moments ($N=0$)

Hence the "safe" condition is OA/OB is less than unity. It can be shown that

$$\frac{\text{OA}}{\text{OB}} = \sqrt{\left(\frac{M_x}{M_{xu}}\right)^2 + \left(\frac{M_y}{M_{yu}}\right)^2} \tag{7.39}$$

where M_{xB}, M_{yB} = limiting values of biaxial moments which together will cause failure,

M_{xu}, M_{yu} = ultimate moments for uniaxial bending only (i.e. zero axial load).

Equation 7.38 also gives the solution for the case of zero axial load (curve EBF in figure 7.15). To obtain the solution for the other end of the interaction curve, the case of axial load only (without moment) is considered. Obviously, failure takes place by buckling about the minor (y–y) axis and the buckling load is $K_{1y}N_u$ (OD in figure 7.15) which can be calculated by the method given in section 7.3.2.

Figure 7.15 shows the interaction of these three parameters (axial load and two moments) for a given column. Curve ED, which is in the plane ($M_y=0$) defines the relation between the axial load and the bending moment about the major axis at failure. Curve FD defines a similar relation for bending about the minor axis alone. When both bending moments are acting, the "safe" space is obtained by rotating DE about the OD axis so that DE contracts to DF while the base travels on the curve EBF.

This three-dimensional diagram can be used to check any combination of the three parameters N, M_x, M_y. For a given set of values, M_x and M_y define the point A and also determine the curve OABD, taking note that the curve OFBE on figure 7.15 is the same as that in figure 7.16. The point H lies on the failure surface and if the acting load N is less than AH, no failure is indicated.

Using the simplified method of section 7.4.1 (equation 7.32), the shape of DHB can be represented by two straight lines as in figure 7.17, which has non-dimensional scale. The procedure of taking the given set N, M_x, M_y is now summarized below, diagram construction being unnecessary (see example 8.4.4). Another advantage of this simplified method, as can be seen below, is to give the designer a better appreciation of the physical implications.

(1) Check if the two moments alone will cause failure ($m > 1$)

$$m^2 = \left(\frac{M_x}{M_{xu}}\right)^2 + \left(\frac{M_y}{M_{yu}}\right)^2 \qquad (7.40)$$

(2) Calculate the concrete contribution factor α from equation 7.5, the squash load N_u from equation 7.4 and K_1 (for the minor axis) from section 7.3.2.
(3) The maximum axial load coexisting with M_x and M_y without causing failure of the column is

$$N_{\max} = N_u[K_1 - m(K_1 - \alpha)] \qquad (7.41)$$

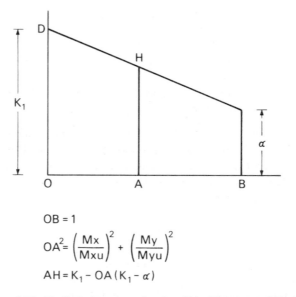

OB = 1

$$OA^2 = \left(\frac{Mx}{Mxu}\right)^2 + \left(\frac{My}{Myu}\right)^2$$

$$AH = K_1 - OA(K_1 - \alpha)$$

Figure 7.17   Vertical plane through point of biaxial moments (M_x, M_y)

7.5   Columns with elastic end restraints

Methods for the assessment of column strength described so far are applicable to pin-ended members only. When there are end restraints such as those due to beams in a framed structure, the assessment of strength would have to take account of the stiffness of the adjoining members. If the beams as well as the column itself become inelastic, ultimate load conditions are difficult to analyse, mainly because hinge rotations of these members are interrelated and their calculation is highly complicated.

When a frame has rigid joints and the adjoining beams can be assumed elastic at column failure, the additional column strength due to the elastic restraint can be allowed for. Application may appear restricted but, in bridges, the use of these hybrid limit states for beam and column is not uncommon, since bridge beams are often designed elastically. This method is referred to as a semi-empirical method in the Bridge Code. It consists of replacing the restrained column by a pin-ended column (with an effective length) subjected to the same load and moments.

The proper treatment of effective length and its widened applications is due to Wood[35] who has published useful charts to enable effective lengths to be determined simply and accurately. His analysis is based on a limited substitute frame consisting of elastic beams of any flexural stiffness, and a column with stiffness related to the axial load. Figures 7.18 and 7.19 show the relevant charts for the "no-sway" and "with sway" conditions respectively.

Having determined the effective length of the equivalent pin-ended column in the plane of bending, the same end load and moments in both planes will be applied and the previous method used to assess the column strength. However, when the column is free to sway, the more critical condition of single-curvature bending must be assumed, irrespective of the signs of the end moments. This implies that if the smaller moment results in double curvature, its sign should be reversed for a subsequent calculation. Furthermore, low values are not recommended, and the Bridge Code requires that the smaller moment should be at least 75% of the larger end moment when the column is free to sway.

7.5.1 *Transverse loads*
These should be included in the elastic analysis of axial loads and moments if their inclusion results in a more severe loading condition. Since all the methods previously described can take account of end moments only, a modified procedure appears reasonable if the moment under the transverse point load (or the maximum moment within the column length) is sufficiently large. The Bridge Code specifies that, for a braced frame, when this maximum moment within the column length exceeds half the modulus (i.e. the numerical value ignoring the sign) of the algebraic sum of the end moments (correct sign of these two moments being used) then an additional analysis must be carried out with the maximum applied at both ends to produce single-curvature bending.

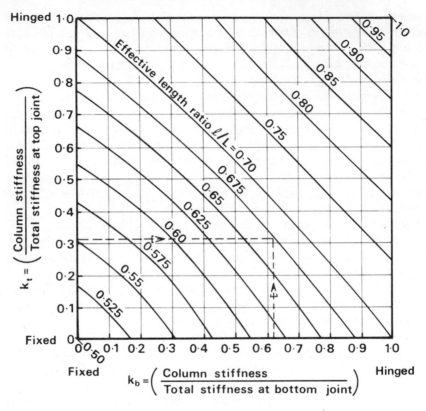

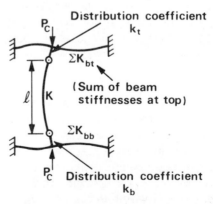

$$k_t = K/(K+\Sigma K_{bt})$$
$$k_b = K/(K+\Sigma K_{bb})$$

Figure 7.18   Effective-length ratios for single column with restraining beams (no sway)

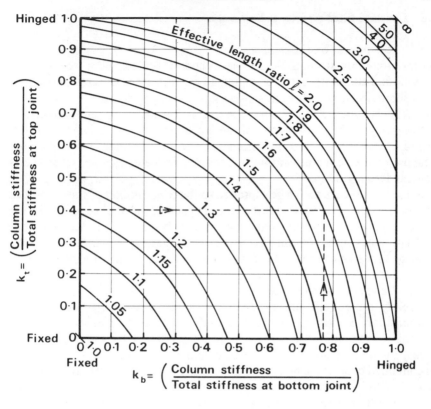

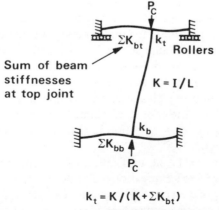

$$k_t = K / (K + \Sigma K_{bt})$$
$$k_b = K / (K + \Sigma K_{bb})$$

Figure 7.19 Effective-length ratios for single column with restraining beams (free to sway)

7.6   Concrete-filled circular hollow sections

7.6.1   *Rectangular and circular sections*

As far as column strength is concerned, concrete-filled hollow sections can provide additional strength due mainly to the restrained concrete, especially with short circular columns in which significant hoop stresses are developed by the steel tubes. Furthermore, the use of formwork or reinforcement is unnecessary in the construction of these columns. One disadvantage is low fire resistance, which makes them more suitable for bridges than buildings. The use of fire-resistant coating is discussed in Reference 36, which gives general design guidance.

Some special care has to be exercised in the construction of concrete-filled hollow sections. Because subsequent inspection of the concrete is not possible, proper compacting has to be ensured. It is also advisable to provide drainage holes at the column base to prevent damage by expansion due to the freezing of water and to vent excessive pressure due to fire or explosion.

7.6.2   *Design of concrete-filled circular columns*

The early theoretical and experimental work on concrete-filled circular columns carried out by Neogi, Sen and Chapman[37] showed that their structural behaviour could reasonably be predicted by the conventional theory for slender columns. In stocky columns, the high steel stresses result in contained concrete subjected to triaxial stresses. Since restrained concrete is known to have higher strength than the "free" concrete in a reinforced concrete column or a cylinder, an enhanced concrete strength should be used for the stocky column.

From a study of the relevant test results, Sen[38] derived an expression for the squash load for concrete-filled circular columns, taking account of the increase in concrete strength due to containment, and the decrease in yield stress of steel due to the particular state of plastic deformation. This plastic deformation consists of a compressive strain in the axial direction and a circumferential tensile strain due to the larger expansion of the concrete. A simple method of evaluating plastic stresses from strain ratios has been developed by Yam[39]; for the case of perpendicular tensile and compressive strain, the yield stress in either direction is always below the uniaxial yield stress assumed in design.

As a further step towards the unification of column design, Virdi and Dowling have reshaped Sen's expression to fit into the slenderness parameter approach described previously. The previous calculation of the inelastic buckling load (for pin-ended columns loaded axially) involves only the elastic flexural stiffness, the squash load, and the standard column curve (*a*, *b* or *c*). In the case of concrete-filled tubes, since the elastic stiffness is unaffected by the containment of concrete, only two problems remain to be solved in order to validate the application of the unified method—to calculate the enhanced squash load and to relate it to slenderness.

Their solution was to modify the strengths of the steel and concrete by some coefficients and relate the coefficients to the slenderness. The contained-concrete strength and the reduced yield stress of the steel casing are respectively:

$$f_{cc} = f_c + c_1 \frac{t}{D} f_s \qquad (7.42)$$

$$f_s' = C_2 f_s \qquad (7.43)$$

where t = wall thickness of steel casing,
        D = outside diameter of column,
   C_1, C_2 = coefficients.

The ratio of column length to diameter (l/D) is then used to vary C_1 and

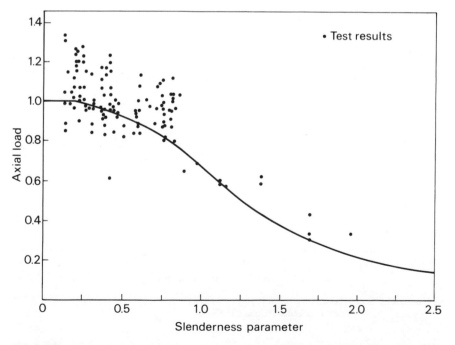

Figure 7.20   Comparison between test results of concrete-filled circular columns and ECCS column curve *a*, including triaxial effects

C_2, and no enhancement is to be considered if the ratio is above 25. Two further coefficients are introduced here for this purpose:

$$C_3 = 0.25 \, (25 - l/D) \quad 0 < C_3 < 6.25 \tag{7.44}$$

$$C_4 = 0.02 \, (25 - l/D) \quad 0 < C_4 < 0.5 \tag{7.45}$$

with

$$C_1 = 2C_2C_3C_4 \tag{7.46}$$

$$C_2 = 1/\sqrt{1 + C_4 + C_4^2} \tag{7.47}$$

Design strengths based on the above method are compared with test results in figure 7.20. Figure 7.21 shows a similar comparison in which the enhancement of concrete strength is ignored. It can be seen from figure 7.21 that column strength will be grossly underestimated for low values of slenderness.

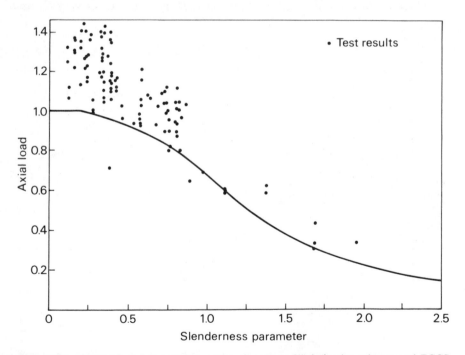

Figure 7.21   Comparison between test results of concrete-filled circular columns and ECCS column curve *a*, ignoring triaxial effects

8   Worked Examples

8.1   Introduction

The purpose of this chapter is to illustrate, by means of numerical examples, the application of some important principles given in previous chapters. The design of the overall structure, and the relevant analysis leading to specific moments and forces on individual members, will not be covered here, and reference to books on structural design and analysis should be made where appropriate. It is essential to refer to the relevant codes of practice, especially in respect of the determination of loads and their combinations. Examples given below will therefore be confined to elements: a simply supported beam, a continuous beam and a column. The first example illustrates the design of simply supported steel beams composite with the concrete floor in a steel-framed building. The design methods comply with CP 117: Part 1: 1965 (see table 2.2) but, in relation to materials and workmanship not covered by CP 117, reference should be made to BS 449 for steel and to CP 114 or CP 110 for concrete. Use should also be made of some helpful publications by Constrado and BCSA on properties of composite sections.[40,41]

The second example is a highway bridge deck with a dual three-lane carriageway plus hard shoulders, and illustrates the design of a continuous composite beam with two equal spans. The cross-section consists of ten plate girders composite with a concrete deck. In order to concentrate on an illustration of some salient features relevant to composite structures, certain parts of the design have been omitted. Thus HA loading (normal traffic) is considered but not HB (abnormal vehicle); design calculations have been carried out for the inner girder only and no analysis made for the transverse load distribution. Other omissions are: stiffening of slender plate components, amount of precambering, fatigue requirements for shear connexion, crack-width control, sequential casting of the slab and transverse bracing to provide stability before the maturing of the concrete. The design calculations comply with the Bridge Code (BS 5400: Part 5) and the appropriate clauses are quoted in the margins of the calculation sheets. It should also be noted that symbols used in this example are identical to those in BS 5400: Part 5.

The final example illustrates the design of a pin-ended composite column subjected to biaxial bending without sway. The collapse criterion of the Bridge Code (BS 5400: Part 5: 1979) is adopted and the calculations developed progressively so as to cover the various cases from pure axial load to biaxial bending. Simplified methods are also given to assist the routine strength checks or assess the adequacy of preliminary designs.

8.2 Design of a simply supported beam

8.2.1 *Design data*

Fixed parameters

Imposed load 5 kN/m^2	Thickness of floor slab 125 mm
Finishes 1 kN/m^2	Dead load due to slab 3 kN/m^2
Span of beams 9 m	Young's modulus for steel 210 kN/mm^2
Spacing of beams 3.5 m	Young's modulus for concrete (210/15) 14 kN/mm^2

Estimated parameters
Dead load due to steel beam 1 kN/m
Flange breadth of steel beam for effective width calculation 150 mm

Parameters to be determined
Steel grade and size of steel beam
Number and spacing of connectors
Amount of transverse reinforcement

8.2.2 *Calculation of loads and load effects*

Working loads

Floor slab	$3 \times 3.5 = 10.50$ kN/m
Steel beam	1.00 kN/m
Construction load (sum of above)	11.50 kN/m
Imposed load + finishes	$(5+1) \times 3.5 = 21.00$ kN/m
Total working load	32.50 kN/m

Bending moments

Due to construction load	$11.5 \times 9^2/8 = 116.4$ kN m
Due to imposed loads	$21 \times 9^2/8 = 212.6$ kN m
Working-load moment	329.0 kN m
Ultimate-load moment	$1.75 \times 329 = 575.8$ kN m

8.2.3 *Effective width*

It should be the least of the following (CP 114, see Table 3.1):

(1) One-third of effective span $9000/3 = 3000$ mm
(2) Beam spacing: 3500 mm
(3) Flange breadth of steel beam plus twelve times the slab thickness $150 + 12 \times 125 = 1650$ mm

Hence the effective width is $b = 1650$ mm
Effective area of concrete slab $A_c = 2063$ cm^2
Moment of inertia $I_c = 269 \times 10^6$ mm^4

8.2.4 Selection of section and checking

Parameters selected
Try 406 × 178 @ 60 Universal Beam (i.e. 0.6 kN/m)

Grade 50: yield strength $f_y = 350$ N/mm^2
Concrete strength 25 N/mm^2 at 28 days $f_{cu} = 25$ N/mm^2

From Table of Properties for Universal Beams

$A_s = 7610$ mm^2 $t = 7.8$ mm
$d = 406.4$ mm $Z_s = 1.059 \times 10^6$ mm^3
$b_f = 177.8$ mm $I_s = 215 \times 10^6$ mm^4
$t_f = 12.8$ mm

Hence $R = f_{cu}/f_y = \frac{4}{9} \times \frac{25}{350} = 0.032$
        $d_c = (406.4 + 125)/2 = 266$ mm

Checking ultimate moment
Case-indicator (figure 3.6):

$$x = \frac{A_s - RA_c}{2b_f} = \frac{7610 - 0.032 \times 206\,300}{2 \times 177.8} = 3 \text{ mm}$$

Since this is less than the flange thickness (12.8 mm), the neutral axis is in the flange and equations 3.27(B) and 3.28(B) apply:

$$n = 125 + 3 = 128 \text{ mm}$$

$$M_p = 350(7610 \times 265.7 - 128 \times 3 \times 177.8) \times 10^{-6} \text{ kN m} = 684 \text{ kN m}$$

Hence the moment capacity exceeds the factored bending moment of 576 kN m.

Checking deflexion
Assuming that the elastic neutral axis is in the steel beam, equation (3.8) is applicable (this equation assumes tensile stiffness of concrete, which is ignored in CP 117) and gives

$$\frac{E_c A_c}{E_s A_s} = \frac{206\,300}{15 \times 7610} = 1.81$$

$$n = \frac{125}{2} + 266 \times \frac{1}{1 + 1.81} = 157 \text{ mm}$$

Since n exceeds 125, the above assumption is correct and from equations 3.11 and 3.12:

$$I_c/15 + I_s = (269/15 + 215) \times 10^6 = 233 \times 10^6 \text{ mm}^4$$

$$\alpha = \frac{1.81}{2.81} \times \frac{7610 \times 266^2}{233 \times 10^6} = 1.48$$

$$M/K = (EI) = 210 \times 233 \times (1 + 1.48) = 122\,000 \text{ kN m}^2$$

Deflexion under live load (21 kN/m) is

$$\delta = \frac{5WL^4}{384(EI)} = \frac{5 \times 21 \times 9^4}{384 \times 122\,000} \times 10^3 = 15 \text{ mm}$$

which is below the limit of 25 mm ($\frac{1}{360}$ of span).

Checking fibre stresses

$$\text{Stress at top of slab} = E_c n k = \frac{M \times n}{(EI)/E_c}$$

$$= \frac{157 \times 329}{122\,000/14} = 6 \text{ N/mm}^2$$

which is less than 8.3 N/mm² (one third of concrete strength).

$$\text{Stress at bottom of steel beam} = \frac{374 \times 329}{122\,000/210} = 212 \text{ N/mm}^2$$

which is less than 315 N/mm² ($0.9 \times$ yield strength).

The composite section is therefore satisfactory. It should also be noted that the estimated parameters (self-weight and flange breadth) err on the safe side and no re-calculation is thus required.

Unpropped construction (for illustration as an alternative)
The steel beam alone will carry the wet concrete plus the self-weight, while the composite beam will be subjected to the additional imposed load. Hence checking involves two components for the steel stress and one for the concrete stress:

$$\text{Concrete fibre stress} = \frac{212.6}{55.3} = 3.8 \text{ N/mm}^2$$

which is less than 8.3 N/mm².

$$\text{Steel fibre stress} = \frac{116.4}{1.059} + \frac{212.6}{1.55} = 247 \text{ N/mm}^2$$

which is less than 315 N/mm².

Design of shear connectors
  Maximum compression in slab $= \frac{4}{9} \times 25 \times 206\,300 \times 10^{-3}$ kN $= 2289$ kN

Try stud connectors 19 mm dia $\times$ 75 mm high.
  From Table 1 of CP 117: Part 1, for the concrete strength of 25 N/mm^2, the design value for a pair of such connectors is 142 kN.

$$\text{Hence number of pairs required} = \frac{2289}{142} = 18 \text{ (say)}$$

Since there are no heavy concentrated loads, these connectors will be spaced evenly between the sections of zero and maximum moment (i.e. over the half span of 4500 mm). The spacing is thus $4500/18 = 250$ mm with end distances of 125 mm.
  It is also necessary to check that the spacing does not exceed four times the slab thickness ($4 \times 125 = 500$ mm) or 600 mm. The present spacing of 250 is well below these limits.

Design of transverse reinforcement
The minimum bar area will be calculated and then two checks made:

$$\text{Minimum area} = \frac{2.5q}{f_{ry}} \text{ cm}^2/\text{m run of beam according to CP 117}$$

$$= \frac{2.5}{250} \times \frac{2289}{4.5} = 5.09 \text{ cm}^2$$

for mild steel bars with a yield strength of 250 N/mm^2
  Hence use 10 mm dia rods at 150 mm centres:

$$A_t = \frac{\pi}{4}(1)^2 \times \frac{1000}{150} = 5.24 \text{ cm}^2$$

The shear resistance in kN/m run should not exceed:

(1) $0.232\,L_s\sqrt{f_{cu}} + 0.1\,A_t f_{ry} n = 290 + 262 = 552$ kN/m

(2) $0.623\,L_s\sqrt{f_{cu}} = 779$ kN/m

where $n =$ number of times each lower bar is intersected by a shear
                surface $= 2$
        $L_s =$ length of shear planes (dotted lines in figure 5.1). Since it should
                not exceed twice the slab thickness, it is taken here as $2 \times 125 =$
                250 mm.

$$q = \frac{2289}{4.5} = 509 \text{ kN/m}$$

Hence the chosen A_t is satisfactory since q is less than 552 and 779 respectively. It should be noted that some longitudinal bars may also be effective in resisting shear.

Bond and anchorage
Any bar in tension should have a sufficient length so that the average bond stress is not excessive. This length is usually expressed as the number of bar diameters and depends mainly on the tensile stress in the bar and the permissible bond stress. For the present example and assuming the maximum tensile strength of 250 N/mm² to be developed, the length required is about 400 mm on each side of the beam centre-line (according to CP 114).

To provide adequate anchorage, the bars should extend at least twelve diameters beyond the points at which they are no longer required to resist stress. For further information, reference to CP 114 or CP 110 should be made.

8.3   Design of a plate girder highway bridge

All symbols refer to BS 5400 (Part 5). References to part 5 are given in the left-hand margin. SLS—Serviceability Limit State, ULS—Ultimate Limit State.

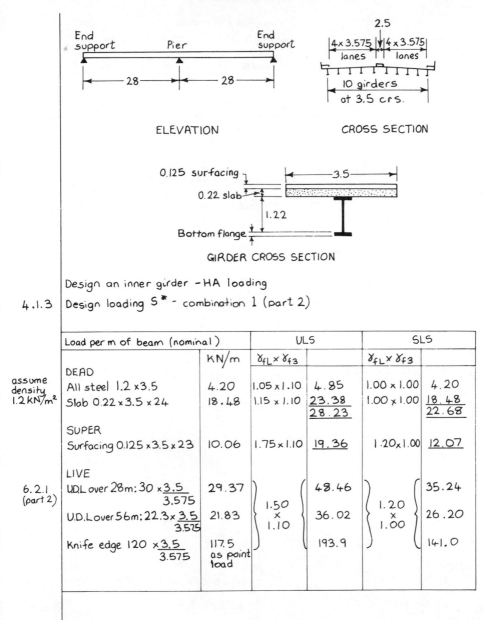

ELEVATION

CROSS SECTION

GIRDER CROSS SECTION

Design an inner girder — HA loading

4.1.3 | Design loading S^* - combination 1 (part 2)

Load per m of beam (nominal)		KN/m	ULS		SLS	
			$\gamma_{fL} \times \gamma_{f3}$		$\gamma_{fL} \times \gamma_{f3}$	
DEAD						
	All steel 1.2 × 3.5	4.20	1.05 × 1.10	4.85	1.00 × 1.00	4.20
	Slab 0.22 × 3.5 × 24	18.48	1.15 × 1.10	23.38	1.00 × 1.00	18.48
				28.23		22.68
SUPER						
	Surfacing 0.125 × 3.5 × 23	10.06	1.75 × 1.10	19.36	1.20 × 1.00	12.07
LIVE						
	U.D.L over 28m: $30 \times \frac{3.5}{3.575}$	29.37		48.46		35.24
	U.D.L over 56m: $22.3 \times \frac{3.5}{3.575}$	21.83	1.50 × 1.10	36.02	1.20 × 1.00	26.20
	Knife edge $120 \times \frac{3.5}{3.575}$	117.5 as point load		193.9		141.0

assume density 1.2 kN/m²

6.2.1 (part 2)

6.1.4.1

<u>BENDING MOMENTS</u> — assuming EI constant
(preliminary)   (i.e. uncracked and shear lag neglected)

DEAD + SUPER + LIVE , both spans , for max. support moment – ULS

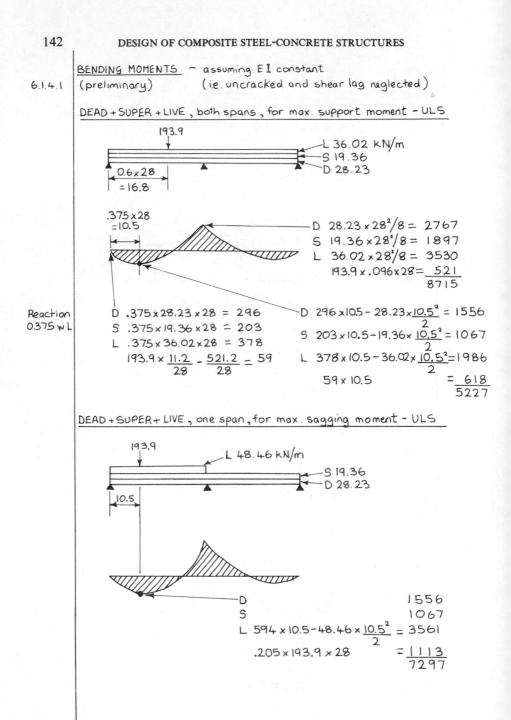

193.9

L 36.02 kN/m
S 19.36
D 28.23

0.6 × 28
= 16.8

.375 × 28
= 10.5

D $28.23 × 28^2/8 = 2767$
S $19.36 × 28^2/8 = 1897$
L $36.02 × 28^2/8 = 3530$
$193.9 × .096 × 28 = \underline{521}$
                    8715

Reaction
0.375 w L

D $.375 × 28.23 × 28 = 296$
S $.375 × 19.36 × 28 = 203$
L $.375 × 36.02 × 28 = 378$
$193.9 × \dfrac{11.2}{28} - \dfrac{521.2}{28} = 59$

D $296 × 10.5 - 28.23 × \dfrac{10.5^2}{2} = 1556$
S $203 × 10.5 - 19.36 × \dfrac{10.5^2}{2} = 1067$
L $378 × 10.5 - 36.02 × \dfrac{10.5^2}{2} = 1986$
$59 × 10.5 \quad = \underline{618}$
                            5227

DEAD + SUPER + LIVE , one span , for max. sagging moment – ULS

193.9

L 48.46 kN/m
S 19.36
D 28.23

10.5

D   1556
S   1067
L $594 × 10.5 - 48.46 × \dfrac{10.5^2}{2} = 3561$
$.205 × 193.9 × 28 \quad = \underline{1113}$
                                7297

TRIAL SECTIONS

Assume live load distribution factor 0.65 which would be confirmed from a distribution analysis.

Materials :   Concrete $f_{cu} = 30$ N/mm^2
              Steel $f_y = 355$ N/mm^2 (grade 50 to B.S. 4360)

PIER

$$\begin{array}{lll} \text{B.M.} & D & 2767 \\ & S & 1897 \\ L\ .65(3530+521)= & \underline{2633} \\ & & 7297 \ \ \text{ULS - Hogging} \end{array}$$

| 4.2.1 Table 1 | For steel $\gamma m = 1.10$ |

Bottom flange area $\dfrac{\text{say } 7297 \times 1.10 \ (\gamma m)}{1.3 \text{ depth} \times 355 \text{ N/mm}^2 \times 10^3} = .0174 \text{m}^2$

                              Try - 500 × 35 plate
                    Try top flange say - 500 × 25 plate

SPAN

$$\begin{array}{lll} \text{B.M.} & D & 1556 \\ & S & 1067 \\ L\ .65(3561+1113)= & \underline{3038} \\ & & 5661 \ \ \text{ULS - Sagging} \end{array}$$

        Bottom flange area $\dfrac{5661 \times 1.10 \times 0.8}{1.3 \times 355 \times 10^3}$ (for composite action)

                                   = .0108 m^2
                                   500 × 22 plate

| Unpropped | Top flange area $\dfrac{(1556+1067) \times 1.10}{1.2 \times 355 \times 10^3} = .0068 \text{m}^2$ |

              (Assume live load   try 400 × 15 plate
              gives nil stress)

| 6.2.4.2 | Assume 10mm web throughout, which must be checked to part 3 assuming steel section alone carries the vertical shear. |

EFFECTIVE BREADTH

| 5.2.3 | For analysis of cross sections |
| 6.1.4.1 | For structural analysis shear lag may be neglected |

| 5.2.3.1 | PIER |

     $\dfrac{b}{l} = \dfrac{1.75}{28} = 0.0625$ $\psi = 0.5375$ (table 4)

| 5.2.3.7 | when cracked $\psi = \psi + \left(\dfrac{1-\psi}{3}\right) = 0.692$ |

SPAN

| 5.2.3.5 | $\dfrac{b}{l} = \dfrac{1.75}{.9 \times 28} = 0.069$ $\psi = 0.969$ mid-span |

                                   $\psi = 0.961$ ¼ span
                                   $\psi = 0.787$ end

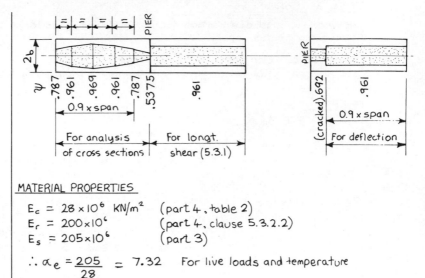

MATERIAL PROPERTIES

$E_c = 28 \times 10^6 \text{ KN/m}^2$ (part 4, table 2)
$E_r = 200 \times 10^6$ (part 4, clause 5.3.2.2)
$E_s = 205 \times 10^6$ (part 3)

$\therefore \alpha_e = \dfrac{205}{28} = 7.32$ For live loads and temperature

$ = \dfrac{205}{23 \times 2} = 14.64$ For super dead loads and shrinkage and deflections.

SECTION PROPERTIES

PIER

		A	$\bar{Y}$ btm	I Steel units	ψ
	Steel	.0422	.569	.01302	–
5.2.3.7	Cracked	.0520	.722	.01822	.692
5.2.3.1	Uncracked $\alpha_e = 7.32$	.0987	1.025	.02836	.5375
5.2.3.1	Uncracked $\alpha = 14.64$	.0704	.889	.02384	.5375

(composite)

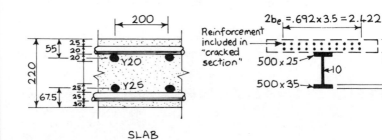

SLAB

SECTION PROPERTIES

SPAN   The critical section is 10.5m from end and ¼-span value
of ψ is used.

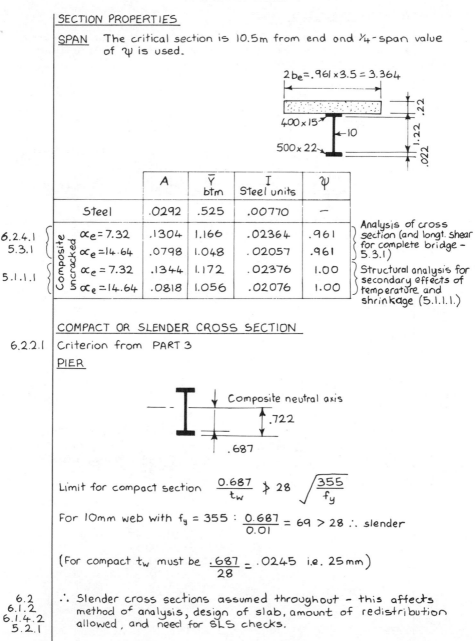

	A	$\bar{Y}$ btm	I Steel units	ψ
Steel	.0292	.525	.00770	—
$\alpha_e = 7.32$	.1304	1.166	.02364	.961
$\alpha_e = 14.64$	.0798	1.048	.02057	.961
$\alpha_e = 7.32$	.1344	1.172	.02376	1.00
$\alpha_e = 14.64$	.0818	1.056	.02076	1.00

6.2.4.1
5.3.1 — Composite uncracked

5.1.1.1

Analysis of cross section (and longt. shear for complete bridge – 5.3.1)

Structural analysis for secondary effects of temperature and shrinkage (5.1.1.1.)

COMPACT OR SLENDER CROSS SECTION

6.2.2.1 Criterion from PART 3

PIER

Composite neutral axis
.722
.687

Limit for compact section $\dfrac{0.687}{t_w} \not> 28\sqrt{\dfrac{355}{f_y}}$

For 10mm web with $f_y = 355$: $\dfrac{0.687}{0.01} = 69 > 28$ ∴ slender

(For compact t_w must be $\dfrac{.687}{28} = .0245$ i.e. 25mm)

6.2
6.1.2
6.1.4.2
5.2.1

∴ Slender cross sections assumed throughout – this affects
method of analysis, design of slab, amount of redistribution
allowed, and need for SLS checks.

BENDING STRESSES   ULS

Bending moments are rechecked taking account of variable inertia, but assuming slab uncracked. Shear lag taken into account (although may be neglected.)

6.1.4.1

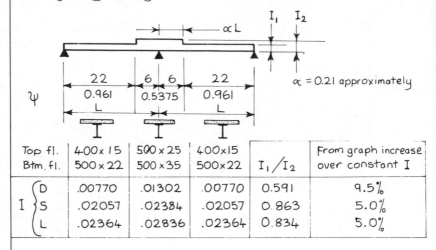

Top fl. Btm. fl.	400 x 15 500 x 22	500 x 25 500 x 35	400 x 15 500 x 22	I_1/I_2	From graph increase over constant I
D	.00770	.01302	.00770	0.591	9.5%
I S	.02057	.02384	.02057	0.863	5.0%
L	.02364	.02836	.02364	0.834	5.0%

BENDING MOMENTS   —   Variable inertia . U.L.S.

Distribution factor 0.65 for live loads.

D+S+L (two spans)

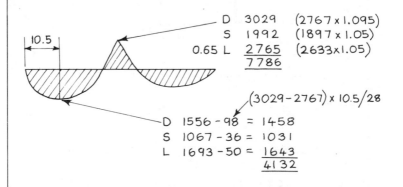

D	3029	(2767 x 1.095)
S	1992	(1897 x 1.05)
0.65 L	2765	(2633 x 1.05)
	7786	

$(3029-2767) \times 10.5/28$

D	1556 - 98 = 1458
S	1067 - 36 = 1031
L	1693 - 50 = 1643
	4132

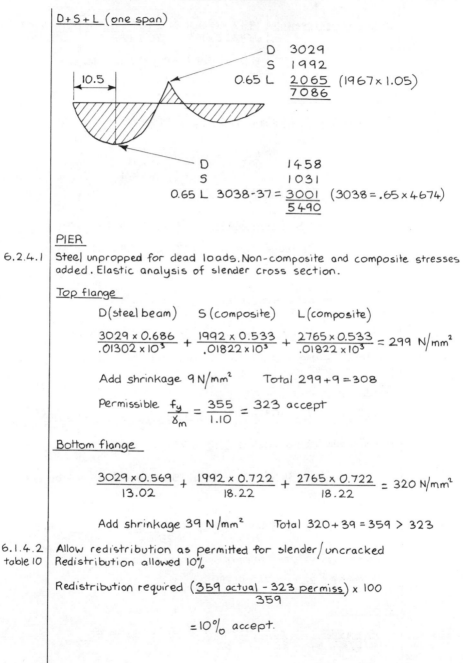

$\underline{D+S+L \ (\text{one span})}$

D   3029
S   1992
0.65 L $\underline{2065}$ (1967×1.05)
7086

D   1458
S   1031
0.65 L   3038-37 = $\underline{3001}$ $(3038 = .65 \times 4674)$
5490

PIER

6.2.4.1   Steel unpropped for dead loads. Non-composite and composite stresses added. Elastic analysis of slender cross section.

Top flange

D(steel beam)   S (composite)   L (composite)

$$\frac{3029 \times 0.686}{.01302 \times 10^3} + \frac{1992 \times 0.533}{.01822 \times 10^3} + \frac{2765 \times 0.533}{.01822 \times 10^3} = 299 \ \text{N/mm}^2$$

Add shrinkage 9 N/mm^2 Total 299+9 = 308

Permissible $\dfrac{f_y}{\gamma_m} = \dfrac{355}{1.10} = 323$ accept

Bottom flange

$$\frac{3029 \times 0.569}{13.02} + \frac{1992 \times 0.722}{18.22} + \frac{2765 \times 0.722}{18.22} = 320 \ \text{N/mm}^2$$

Add shrinkage 39 N/mm^2 Total 320+39 = 359 > 323

6.1.4.2   Allow redistribution as permitted for slender/uncracked
table 10   Redistribution allowed 10%

Redistribution required $\dfrac{(359 \ \text{actual} - 323 \ \text{permiss})}{359} \times 100$

$= 10\%$ accept.

<u>Bottom reinforcement</u> $\left[\dfrac{1992 \times 0.601}{0.01822 \times 10^3} + \dfrac{2765 \times 0.601}{0.01822 \times 10^3}\right] \dfrac{200 \times 10^6}{205 \times 10^6}$

$$\left[\underset{\substack{\text{super} \\ \text{dead}}}{64} + \underset{\text{live}}{89}\right] = \underline{153\,N/mm^2}\ \text{tension}$$

<u>Top reinforcement</u>

Add shrinkage 44   Total 178 + 44 = 222 $\underline{178\,N/mm^2}$ tension

$$\not> \dfrac{460}{1.15} = 400\ N/mm^2$$
$$-\ accept$$

<u>SPAN</u> (10.5 m from end)

$\quad$ <u>D + S + L (both spans)</u>   Moment = 4132 KNm

$\quad$ Add 10% redistribution : 10% × $(7786 + 1117) = 890$
$$\qquad\qquad\qquad\qquad\quad \uparrow \qquad\quad \uparrow$$
$$\qquad\qquad\qquad\qquad D+S+L \quad \substack{secondary \\ shrinkage}$$

$\quad$ Total BM = 4132 + 890 × $\dfrac{10.5}{28}$ = 4466

<u>D + S + L (one span)</u>   1458 + 1031 + 3001 = 5490 KNm

6.1.4.2 For this loading, redistribution has not been used to satisfy stress at pier.

<u>Top flange</u>

$\quad \dfrac{1458 \times 0.717}{7.7} + \dfrac{1031 \times 0.194}{20.57} + \dfrac{3001 \times 0.076}{23.64} = \dfrac{156\,N/mm^2}{compression}$

$\quad$ For grade 43 permissible $\dfrac{250}{1.1}$ = 227 accept.

$\quad$ Add shrinkage 34 N/mm²   Total 156 + 34 = $\underline{190\,N/mm^2}$ accept.

<u>Bottom flange</u>

$\quad \dfrac{1458 \times 0.525}{7.7} + \dfrac{1031 \times 1.048}{20.57} + \dfrac{3001 \times 1.166}{23.64} = \dfrac{300\,N/mm^2}{tension}$

$\quad$ (Shrinkage 15 N/mm² compression, hence not added)

$\quad$ Permissible $\dfrac{355}{1.10}$ = 323 accept.

<u>Top of slab</u>

$\quad \dfrac{1031 \times 0.414}{20.57 \times 14.64} + \dfrac{3001 \times 0.296}{23.64 \times 7.32} = 1.42 + 5.13 = \underline{6.55}$

LOCAL SLAB BENDING (longitudinal)

6.2.5 (Part 2) Due to 100 KN single wheel load - this load is alternative to HA -(UDL + knife edge load)

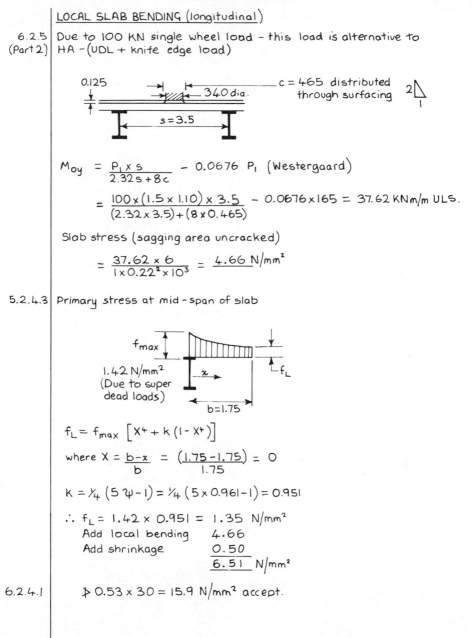

$$M_{oy} = \frac{P_1 \times s}{2.32s + 8c} - 0.0676 \, P_1 \quad \text{(Westergaard)}$$

$$= \frac{100 \times (1.5 \times 1.10) \times 3.5}{(2.32 \times 3.5) + (8 \times 0.465)} - 0.0676 \times 165 = 37.62 \, \text{KNm/m ULS}.$$

Slab stress (sagging area uncracked)

$$= \frac{37.62 \times 6}{1 \times 0.22^2 \times 10^3} = \underline{4.66} \, \text{N/mm}^2$$

5.2.4.3 Primary stress at mid-span of slab

$$f_L = f_{max} \left[X^+ + k (1 - X^+) \right]$$

where $X = \dfrac{b-x}{b} = \dfrac{(1.75 - 1.75)}{1.75} = 0$

$k = \frac{1}{4} (5\psi - 1) = \frac{1}{4} (5 \times 0.961 - 1) = 0.951$

$\therefore f_L = 1.42 \times 0.951 = 1.35 \, \text{N/mm}^2$

Add local bending   4.66
Add shrinkage $\underline{0.50}$
                   $\underline{6.51}$ N/mm²

6.2.4.1 $\not> 0.53 \times 30 = 15.9 \, \text{N/mm}^2$ accept.

PIER – SLAB CRACKED   Local bending 37.62 KN m/m

152.5

N.A. d_n

Soffit→

67.5

25 at 200: 2453 mm²/m

$$\frac{E_s}{E_c} = \frac{200}{28} = 7.14$$

$$2453(152.5 - d_n) = \frac{1000 \times d_n^2}{7.14 \times 2}$$

$\therefore d_n = 57.64$ $I = 0.3101 \times 10^{-4}$ reinforcement units

conc. stress $\dfrac{37.62 \times 0.0576}{0.3101 \times 10^{-4} \times 7.14 \times 10^{3}} = 9.79$ N/mm² compression

reinft - stress $\dfrac{37.62 \times 0.0949}{0.3101 \times 10^{-4} \times 10^{3}} = 115$ N/mm² tension

$f_{max} = 64$
super
dead

f_L

$K = \frac{1}{4}(5 \times 0.692 - 1) = 0.615$

$\therefore f_L = 64 \times 0.615 = 39$

add local bending $\dfrac{115}{154}$ N/mm²

Add shrinkage 67 N/mm²   Total 221 N/mm²   Permissible $\dfrac{425}{1.15}$

5.4.3   SHRINKAGE AND CREEP

4.1.3   Partial safety factors $\delta_{fL} \times \delta_{f3} = 1.2 \times 1.1 = \underline{1.32}$

Table 9   Shrinkage strain (general open air) $= \underline{200 \times 10^{-6}}$

Modified modular ratio (due to creep)

$$\alpha_e = \frac{205 \times 10^{6}}{14 \times 10^{6}} = \underline{14.64}$$

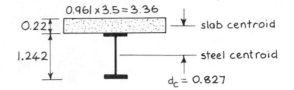

For primary effects the transformed section is used and effective breadth at quarter span adopted ($\psi = 0.961$, hence $I = 0.02057$ from table of section properties for "span")

$A_s = 0.0292$
$A_c = 0.74$
$E_s / E_c = 14.64$
$I_s = 0.0077$
$I_c = 0.0030$

Section moduli

$Z \text{(slab)} = 0.027$

$Z_{s3} = \dfrac{0.0077}{0.717} = 0.0107$

$Z_{s4} = \dfrac{0.0077}{0.525} = 0.0147$

Curvature and slab force due to shrinkage are calculated from equations 6.12 and 6.13.

Composite stiffness factor (α)

$$\frac{A_c \times \alpha_e A_s}{A_c + \alpha_e A_s} \times \frac{d_c^2}{I_c + \alpha_e I_s}$$

$$= \frac{0.74 \times 14.64 \times 0.0292}{0.74 + 14.64 \times 0.0292} \times \frac{0.827^2}{0.0030 + 14.64 \times 0.0077}$$

$$= \underline{1.615} \qquad \text{Equation (3.12)}$$

Curvature $K = \dfrac{\alpha (\delta_{fL} \times \delta_{f3} \times e_1)}{(1 + \alpha) d_c}$ \qquad Equation (6.12)

$$= \frac{1.615 (1.32 \times 200 \times 10^{-6})}{2.615 \times 0.827} = \underline{197 \times 10^{-6} \text{ m}^{-1}}$$

Slab force $= \dfrac{E_s A_s A_c}{A_c + \alpha_e A_s} \times \dfrac{\delta_{fL} \, \delta_{f3} \, e_1}{1 + \alpha}$ \qquad Equation (6.13)

$$= \frac{205 \times 10^6 \times 0.0292 \times 0.74}{0.74 + 14.64 \times 0.0292} \times \frac{264 \times 10^{-6}}{2.615} = \underline{386 \text{ KN}}$$

<u>Shrinkage stresses</u> \qquad Equations (6.14 – 6.17)

<u>SPAN</u>

Top of slab

$$\left[-\frac{386}{0.74}+\frac{197\times14\times0.003}{0.027}\right]\frac{N/mm^2}{10^3}=-0.2$$

Bottom of slab

$$\left[-\frac{386}{0.74}-\frac{197\times14\times0.003}{0.027}\right]\frac{N/mm^2}{10^3}=-0.8$$

Top of steel beam

$$\left[\frac{386}{0.0292}+\frac{197\times205\times0.0077}{0.0107}\right]\frac{N/mm^2}{10^3}=42.3$$

Bottom of steel beam $$\left[\frac{386}{0.0292}-\frac{197\times205\times0.0077}{0.0147}\right]\frac{N/mm^2}{10^3}=-8$$

Secondary effects are due to constant curvature ($k=197\times10^{-6}$) which induces reactions at supports.

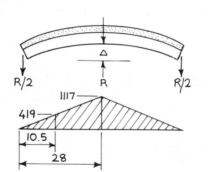

For constant k

$$\Delta=\frac{k}{8}(span)^2$$

$$=\frac{197\times10^{-6}\times56^2}{8}=0.07m$$

Reaction required to produce Δ :

$$R=\frac{48(EI)\Delta}{(span)^3}$$

5.4.2.1
5.1.1.1

Use uncracked section with $\psi=1$

$$R=\frac{48\times205\times10^6\times0.02076\times0.07}{56^3}$$

$$=80\ KN$$

Moment at pier $=\dfrac{R}{2}\times28=\underline{1117KNm}$

<u>Additional stresses</u> (secondary effects)

5.4.2.1 Use uncracked section with $\psi=0.961$ $(I=0.02057)$

5.2.3 B.M. at section 10.5m from end $= 419$ KNm (hogging)

Top of slab

$$-\frac{419\times0.414}{0.02057\times14.64\times10^3}=-0.6\ N/mm^2$$

Bottom of slab

$$-\frac{419\times0.194}{20.57\times14.64}=-0.3\ N/mm^2$$

Top of steel beam $-\dfrac{419 \times 0.194}{20.57} = -4.0 \text{ N/mm}^2$

Bottom of steel beam $\dfrac{419 \times 1.048}{20.57} = 21.3 \text{ N/mm}^2$

Hence total stresses are :

$-0.8, -1.1, 38$ and 13 respectively

PIER

5.4.2.1
Table 8

Uncracked section assumed except for longitudinal bending stresses due to secondary effects i.e. primary stresses and reaction at pier support are based on uncracked section while bending stresses due to induced reaction based on cracked section with $\psi = 0.692$.

Results are summarized below.

6.1.5

TEMPERATURE   ULS considered for slender section

$\gamma_{fL} \times \gamma_{f3} = 1.0 \times 1.1 = 1.1$ Combination 3 (Part 2)

Calculation of stresses due to temperature difference through depth (primary effects) and induced by support reaction (secondary effects) is similar to shrinkage analysis.

$\alpha_e = 7.32$ (short-term)

5.4.3
Part 2

Temperature difference from BS 5400 (Table 12 and Figure 9 of Part 2). Stresses from equations 6.5 and 6.6 or 6.14 – 6.19 using an equivalent slab temperature (constant through slab depth). Results based on equivalent temperature (positive) summarized below.

SUMMARY OF STRESSES   ULS

(Excluding local slab bending)

COMBINATION 1

PIER

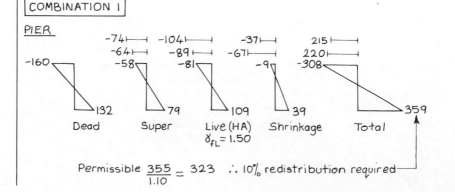

Permissible $\dfrac{355}{1.10} = 323$ ∴ 10% redistribution required

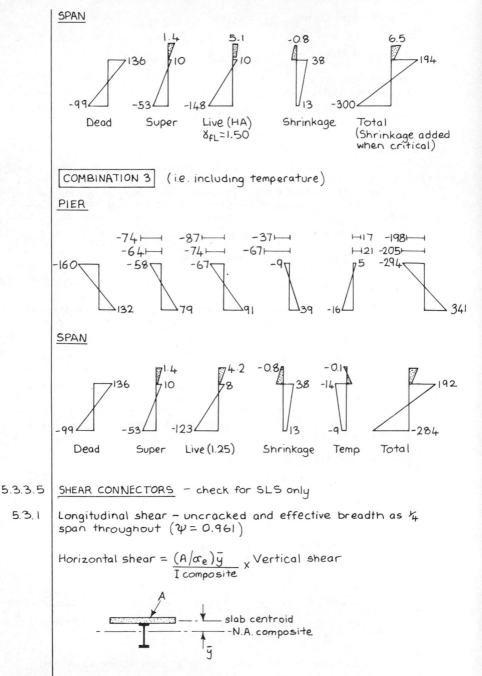

SPAN

1.4	5.1	-0.8	6.5
136 →10	→10	38	→194
-99 -53	-148	13 -300	
Dead	Super Live (HA) $\gamma_{fL}=1.50$	Shrinkage	Total (Shrinkage added when critical)

COMBINATION 3   (i.e. including temperature)

PIER

-74 ⊢⊣	-87 ⊢⊣	-37 ⊢⊣	⊢17 -198⊢⊣
-64 ⊢⊣	-74 ⊢⊣	-67 ⊢⊣	⊢21 -205⊢⊣
-58	-67	-9	5 -294
-160			
132	79	91	39 -16
			341

SPAN

1.4	4.2 -0.8	-0.1	192
136 →10	→8	38 -14	
-99 -53	-123	13 -9	-284
Dead	Super Live (1.25)	Shrinkage Temp	Total

5.3.3.5 SHEAR CONNECTORS – check for SLS only

5.3.1 Longitudinal shear – uncracked and effective breadth as $\frac{1}{4}$ span throughout ($\psi = 0.961$)

Horizontal shear $= \dfrac{(A/\alpha_e)\bar{y}}{I\ \text{composite}} \times$ Vertical shear

A

slab centroid

N.A. composite

$\bar{y}$

For $\alpha_e = 14.64$ (super dead): $\dfrac{0.0506 \times 0.304}{0.02057} = 0.7478$

For $\alpha_e = 7.32$ (live): $\dfrac{0.1012 \times 0.186}{0.02364} = 0.7964$

Assume distribution factor for live U.D.L. $= 0.65$ but assume 1.0 for K.E. load because no distribution will occur when K.E. load is at supports.

<u>SHEAR ENVELOPES</u> — vertical (horizontal shear in parentheses)

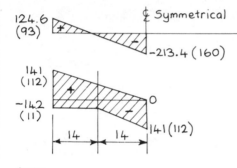

124.6
(93)

−213.4 (160)

¢ Symmetrical

Super dead
$(\gamma_{fL} \times \gamma_{f3} = 1.20 \times 1.00)$

141
(112)

−142
(11)

14 14

141(112)

O

Live K.E.
(1.20×1.00)

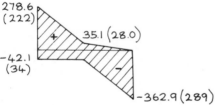

278.6
(222)

35.1 (28.0)

−42.1
(34)

−362.9 (289)

Live U.D.L.
(1.20×1.00)

| 5.4.3 | <u>SHRINKAGE</u> $(\gamma_{fL} \times \gamma_{f3} = 1.00 \times 1.00)$ $\alpha_e = 14.64$ |

5.4.2.3 Transfer length $l_s = 2\sqrt{\dfrac{kQ}{\Delta f}}$

$Q = 386/1.32 = 291.8$ KN
$K = 0.003$ studs
$\Delta f = \varepsilon_{cs} = 200 \times 10^{-6}$

$$= 2\sqrt{\dfrac{0.003 \times 291.8}{200 \times 10^{-6} \times 10^3}} = \underline{4.184 \text{ m}}$$

Horizontal shear at end $= \dfrac{2Q}{l_s} = \dfrac{2 \times 291.8}{4.184} = \underline{140 \text{ KN/m}}$

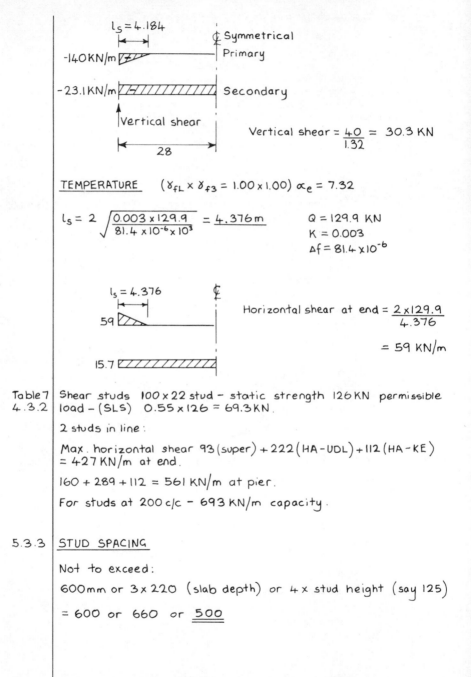

$l_s = 4.184$

℄ Symmetrical
Primary

-140 KN/m

-23.1 KN/m   Secondary

Vertical shear

28

Vertical shear = $\dfrac{40}{1.32}$ = 30.3 KN

TEMPERATURE $(\gamma_{fL} \times \gamma_{f3} = 1.00 \times 1.00)\ \alpha_e = 7.32$

$l_s = 2\sqrt{\dfrac{0.003 \times 129.9}{81.4 \times 10^{-6} \times 10^3}} = 4.376\,m$

$Q = 129.9$ KN
$K = 0.003$
$Af = 81.4 \times 10^{-6}$

$l_s = 4.376$

℄

59

15.7

Horizontal shear at end = $\dfrac{2 \times 129.9}{4.376}$

= 59 KN/m

| Table 7 | Shear studs 100 x 22 stud – static strength 126 KN permissible |
| 4.3.2 | load – (SLS) 0.55 × 126 = 69.3 KN. |

2 studs in line:

Max. horizontal shear 93 (super) + 222 (HA-UDL) + 112 (HA-KE)
= 427 KN/m at end.

160 + 289 + 112 = 561 KN/m at pier.

For studs at 200 c/c – 693 KN/m capacity.

5.3.3   STUD SPACING

Not to exceed:

600mm or 3 x 220 (slab depth) or 4 x stud height (say 125)

= 600 or 660 or 500

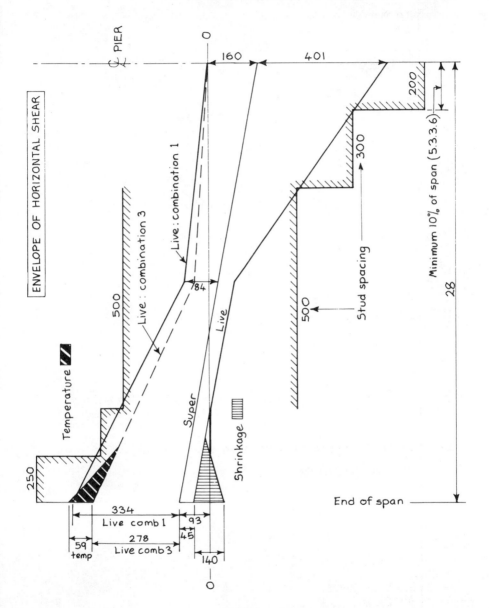

ENVELOPE OF HORIZONTAL SHEAR

Transverse reinforcement

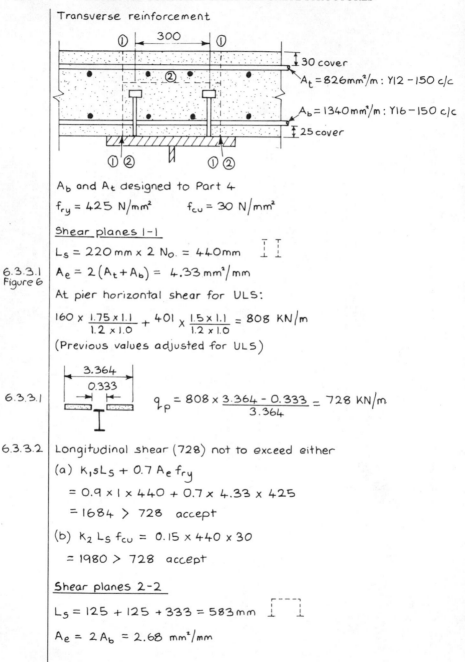

A_b and A_t designed to Part 4

$f_{ry} = 425 \text{ N/mm}^2$ $f_{cu} = 30 \text{ N/mm}^2$

Shear planes 1-1

$L_s = 220 \text{ mm} \times 2 \text{ No.} = 440 \text{mm}$

6.3.3.1
Figure 6

$A_e = 2(A_t + A_b) = 4.33 \text{ mm}^2/\text{mm}$

At pier horizontal shear for ULS:

$$160 \times \frac{1.75 \times 1.1}{1.2 \times 1.0} + 401 \times \frac{1.5 \times 1.1}{1.2 \times 1.0} = 808 \text{ KN/m}$$

(Previous values adjusted for ULS)

6.3.3.1

$$q_p = 808 \times \frac{3.364 - 0.333}{3.364} = 728 \text{ KN/m}$$

6.3.3.2

Longitudinal shear (728) not to exceed either

(a) $K_1 s L_s + 0.7 A_e f_{ry}$

   $= 0.9 \times 1 \times 440 + 0.7 \times 4.33 \times 425$

   $= 1684 > 728$ accept

(b) $K_2 L_s f_{cu} = 0.15 \times 440 \times 30$

   $= 1980 > 728$ accept

Shear planes 2-2

$L_s = 125 + 125 + 333 = 583 \text{mm}$

$A_e = 2A_b = 2.68 \text{ mm}^2/\text{mm}$

$$q_p = 808 \times \frac{3.364 \times 0.22 - 0.333 \times 0.125}{3.364 \times 0.22}$$

$$= 763 \text{ KN/m}$$

(a) $0.9 \times 583 + 0.7 \times 2.68 \times 425$

$= 1322 > 763$ accept

(b) $0.15 \times 583 \times 30$

$= 2624 > 763$ accept

6.3.3.3 Interaction between shear and bending

Sagging — not likely to occur

Hogging — minimum B.M. $= 0.1 \, wL^2$

$= 0.1 \times 14.24 \text{ (dead)} \times 3.5^2 = 17.44 \text{ KNm}$

$\therefore F_T \simeq \dfrac{17.44}{0.189 \times 0.85} = 109 \text{ KN/m}$

Consider shear planes 2-2 only

Criterion: $q_p \leqslant k_1 \, s \, L_s + 0.7 \, A_e \, f_{ry} + 1.6 \, F_T$

$0.9 \times 1 \times 583 + 0.7 \times 2.68 \times 425 + 1.6 \times 109$

$= 1496 \text{ KN/m} > 763$ accept

6.3.3.4 Minimum transverse reinforcement

$$= \frac{0.8 \times s \times h_c}{f_{ry}} = \frac{0.8 \times 1 \times 220}{460}$$

$$= 0.383 \text{ mm}^2/\text{mm} = \underline{383 \text{ mm}^2/\text{m width}}$$

8.4 Design of a composite column

8.4.1 *Design data*

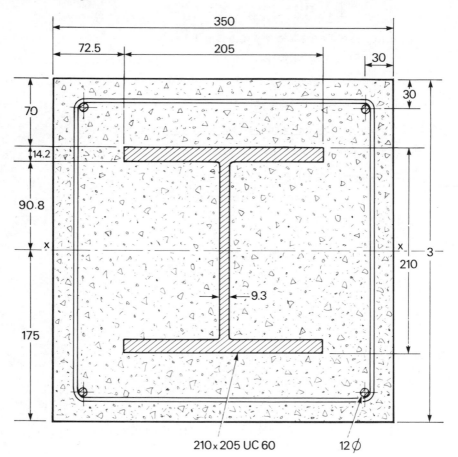

Figure 8.1 Cross-section of composite column

Loading (inclusive of partial load factors)

Axial loads	Dead load	$N_g =$ 700 kN
	Live load	$N_q = \underline{1500}$ kN
	Total load	= 2200 kN
Moments about major axis $(x–x)$	Dead load	$M_g =$ 13 kN m
(including eccentricity)	Live load	$M_q = \underline{30}$ kN m
	Total	= 43 kN m

Moment at lower end produces double curvature, hence use minimum value to obtain more critical loading condition. The dead load moment at the lower end is also 13 kN m, hence:

	$\beta = \frac{13}{43}$	$= -0.3$
Moments about minor axis $(y-y)$	Dead load	$M_g = 8$ kN m
(including eccentricity)	Live load	$M_q = 12$ kN m
	Total	$= 20$ kN m

Bending in single curvature with maximum moment at the lower end of 8 kN m, hence

$$\beta = \tfrac{8}{20} \qquad\qquad = 0.4$$

Properties of column to be checked

Effective length	$l = 4.5$ m
Structural steel: Grade 50	$f_y = 350$ N/mm^2
210×205 UC @ 60 (figure 8.1)	$A_s = 75.8$ cm^2
	$I_{sx} = 6088$ cm^4
	$I_{sy} = 2041$ cm^4
	$E_s = 205$ kN/mm^2
Concrete: 28-day characteristic cube strength	$f_{cu} = 30$ N/mm^2
	$E_c = 13.5$ kN/mm^2
Reinforcement: $4 \times 12\phi$ high-yield bars	$f_{ry} = 410$ N/mm^2
	$A_r = 4.5$ cm^2
	$E_r = 200$ kN/mm^2

8.4.2 Preliminary calculations

Area and moments of inertia

$$A_c = 35 \times 35 - 75.8 - 4.5 = 1145 \text{ cm}^2$$
$$I_r = 4.5(17.5 - 3.0)^2 = 946 \text{ cm}^4$$
$$I_{cx} = 35^4/12 - 6088 - 946 = 1.18 \times 10^5 \text{ cm}^4$$
$$I_{cy} = 35^4/12 - 2041 - 946 = 1.22 \times 10^5 \text{ cm}^4$$

Ultimate moment of resistance (bending only)

$$f_s = 350/1.10 = 318 \text{ N/mm}^2$$
$$f_c = 0.4 \times 30 = 12 \text{ N/mm}^2$$
$$f_r = 410/1.15 = 357 \text{ N/mm}^2$$

STEEL SECTION ONLY

$$M_p = 2 \times 318(205 \times 14.2 \times 97.9 + 9.3 \times 90.8 \times 45.4) \times 10^{-6} \text{ kN m} = 205.6 \text{ kN m}$$

REINFORCEMENT ONLY

$$M_r = 357 \times 450 \times 145 \times 10^{-6} \text{ kN m} = 23.3 \text{ kN m}$$

COMPOSITE SECTION (ABOUT MAJOR AXIS)
Test position of neutral axis, using equation (7.35) and setting $N_3 = 0$, since there is no axial compression:

$$y = \frac{175(2 \times 9.3 \times 318)}{350 \times 12 + 2 \times 9.3 \times 318} = 102 \text{ mm}$$

Hence the neutral axis is in the web as assumed and equations (7.35) and (7.36) are applicable:

$$M_u = [\tfrac{1}{2} \times 350 \times 12 \times 102(350 - 102) - 9.3 \times 318(175 - 102)^2] \times 10^{-6} + M_p + M_r$$
$$= 53.2 - 15.6 + 205.6 + 23.3 = 267 \text{ kN m}$$

COMPOSITE SECTION (ABOUT MINOR AXIS)
The neutral axis is in the flanges:

$$y = \frac{7580 + 4 \times 14.2 \times 72.5}{350 \times 0.0377 + 4 \times 14.2} = 167 \text{ mm}$$

$$M_u = [12 \times 350 \times 167(175 - 83.5) + 318 \times 56.8(167 - 72.5)(175 - 167 + 47.3)] \times 10^{-6} + M_r$$

$$M_u = 182 \text{ kN m}$$

8.4.3   Checking ultimate strength

Axial load only, buckling about major axis only

From BS 5400: Part 5

$$N_u = 0.91 \, A_s f_y + 0.87 \, A_r f_{ry} + 0.45 \, A_c f_{cu}$$
$$= (0.91 \times 75.8 \times 350 + 0.87 \times 4.5 \times 410 + 0.45 \times 1145 \times 30) \times 10^{-1} \text{ kN}$$
$$= 4121 \text{ kN}$$
$$\alpha = 0.45 \, A_c f_{cu}/N_u = 0.375 \quad [\text{see } (7.5)]$$

Since this value lies between 0.15 and 0.80, the methods in 7.3 and 7.4 can be applied. (Another condition is that l/b lies between 12 and 30: $4500/350 = 13$).

$$l_E^2 = \pi^2(13.5 \times 1180 + 205 \times 60.88 + 200 \times 9.46)/4121 \text{ m}^2 \quad [\text{see } (7.13)]$$
$$l_E = 8.52 \text{ m}$$
$$\lambda_x = 4.5/8.52 = 0.53 \quad [\text{see } (7.14)]$$

From Table 13 of the Code (or figure 7.4) it will be seen that the curve *b* is appropriate for this steel section. For $\lambda = 0.53$, Table 13.2 of the Code gives:

$$K_1 = 0.872$$

Hence, if no bending moments are acting, the buckling load is

$$N_{ax} = 0.872 \times 4121 = 3594 \text{ kN}$$

Axial load and uniaxial bending (about major axis)

$$\frac{K_2}{K_{20}} = \frac{90 - 25(-0.6 - 1)(1.8 - 0.375) - 120 \times 0.53}{30(2.5 + 0.3)} = 0.99$$

This lies between 0 and 1 and is therefore acceptable.

$$K_2 = 0.99 \, K_{20} = 0.99 \times [0.9(0.375)^2 + 0.2] = 0.324$$

Since bending about the strong axis is being considered ($\lambda_x < \lambda_y$; see λ_y below), K_3 should be taken as zero.

The maximum safe axial load which can be applied with the moment of magnitude 43 kN m is therefore given by

$$\frac{N}{4121} = 0.872 - (0.872 - 0.324)(43/267) \quad [\text{see } (7.26)]$$

$$N_x = 3230 \text{ kN}$$

which is much higher than the maximum design load (2200 kN).

With biaxial bending

Before applying equation 7.37, the failure load under uniaxial bending about the minor axis should first be calculated.

$$l_E^2 = \pi^2(13.5 \times 1220 + 205 \times 20.41 + 200 \times 9.46)/4121$$
$$l_E = 7.35$$
$$\lambda_y = 4.5/7.35 = 0.61$$

From Tables (curve c applicable)

$$K_1 = 0.776$$

$$\frac{K_2}{K_{20}} = \frac{90 - 25(0.8 - 1)(1.8 - 0.375) - 140 \times 0.61}{30(2.5 - 0.4)} = 0.186$$

$$K_{20} = 0.9 \times 0.375^2 + 0.2 = 0.327$$

$$K_2 = 0.186 \times 0.327 = 0.061$$

$$K_3 = 0.425 - 0.075 \times 0.4 - 0.005 \times 140 \times 0.61 = -0.03$$

which is within the following two limits:

$$-0.03 \times 1.4 = -0.04$$
$$0.2 - 0.25 \times 0.375 = 0.11$$

Hence N_y is given by (equation 7.26):

$$\frac{N_y}{4121} = 0.776 - (0.776 - 0.061 + 4 \times 0.03)\left(\frac{20}{182}\right) + 4 \times 0.03\left(\frac{20}{182}\right)^2$$

$$N_y = 2825 \text{ kN}$$

Hence the failure axial load under biaxial bending is N_{xy}, where

$$\frac{1}{N_{xy}} = \frac{1}{3230} + \frac{1}{2825} - \frac{1}{2594}$$

$$N_{xy} = 2594 \text{ kN}$$

Since the maximum design load (2200 kN) is below this ultimate capacity (2595 kN), the column is satisfactory.

8.4.4 Simplified method

From equation (7.33)

$$N_x = 4121[0.872 - \tfrac{43}{267}(0.872 - 0.375)] = 3264 \text{ kN}$$
$$N_y = 4121[0.776 - \tfrac{20}{182}(0.776 - 0.375)] = 3016 \text{ kN}$$

Hence

$$\frac{1}{N_{xy}} = \frac{1}{3264} + \frac{1}{3016} - \frac{1}{3594}$$

$$N_{xy} = 2779 \text{ kN}$$

These three values differ from the previous ones by 1%, 7% and 7% respectively. An even simpler method is provided by equations (7.40) and (7.41).

Check if the two moments alone will cause failure.

$$m^2 = (\tfrac{43}{267})^2 + (\tfrac{20}{182})^2$$
$$m = 0.195 \quad \text{(less than unity)}$$

Hence $N_{xy} = 4121[0.776 - 0.195(0.776 - 0.375)] = 2876 \text{ kN}$. This differs from the Bridge Code value by 11% but it is worth noting that the latter is based on an approximate formula.

References

(1) Joint ASCE-AASHO Committee on Flexural Members, "Design of hybrid steel beams", *Journal of the Structural Division Proceedings of the American Society of Civil Engineers*, Paper 5995, June 1968, pp 1397–1426.

(2) Basler, K, "Strength of plate girders in shear", *Journal of the Structural Division, ASCE*, Vol 87, No ST7, Proc Paper 2967, October 1961, pp 151–180.

(3) Yam, L C P, "Functional requirements of current structural codes in the UK", *IABSE Colloquium on "Codes of Practice—Reform or Decline"*, Cambridge, July 1979.

(4) Somerville, G, "Background on concrete codes in the UK", *IABSE Colloquium on "Codes of Practice—Reform or Decline"*, Cambridge, July 1979.

(5) Yam, L C P and Chapman, J C, "The inelastic behaviour of simply supported composite beams of steel and concrete", *Proceedings of the Institution of Civil Engineers*, Vol 41, Paper 7111, December 1968, pp 651–683.

(6) Johnson, R P and May, I M, "Partial-interaction design of composite beams", *The Journal of the Institution of Structural Engineers*, Vol 53, No 8, August 1975, pp 305–311.

(7) Johnson, R P, *Composite structures of steel and concrete, Vol 1. Beams, columns, and frames and applications in buildings*, Crosby Lockwood Staples, 1975.

(8) McGarraugh, J B and Baldwin, J W, "Lightweight concrete-on-steel composite beams", *Engineering Journal, American Institute of Steel Construction*, Vol 8, July 1971, p 90.

(9) Frodin, J G, Taylor, R and Stark, J W B, "A comparison of deflexions in composite beams having full and partial shear connexion", *Proceedings of ICE*, Part 2, Vol 65, Paper 8103, June 1978, pp 307–322.

(10) Yam, L C P and Chapman, J C, "The inelastic behaviour of continuous composite beams of steel and concrete", *Proc ICE*, Vol 53, No 7551, December 1972, pp 487–501.

(11) Hope-Gill, M C, "Redistribution in composite beams", *Journal ISE*, Vol 57B, No 1, March 1979, pp 7–10.

(12) Yam, L C P, "Ultimate-load behaviour of composite T-beams having inelastic shear connexion", PhD Thesis, University of London, December 1966.

(13) Hamada, S and Longworth, J, "Ultimate strength of continuous composite beams", *Journal of the Structural Division, ASCE*, Paper 12267, July 1976, pp 1463–1478.

(14) Mallick, S K and Chattopadhyay, S K, "Ultimate strength of continuous composite beams", *Building Science*, Vol 10, Pergamon Press 1975, pp 189–198.

(15) Rotter, J M and Ansourian, P, "Cross-section behaviour and ductility in composite beams", *Proc ICE*, Part 2, Vol 67, Paper 8213, June 1979, pp 453–474.

(16) BS 5400: "Steel, concrete and composite bridges, Part 5: Code of Practice for Design of Composite Bridges", British Standards Institution, 1979.

(17) Evans, R H and Kong, F K, "The extensibility and micro-cracking of the in-situ concrete in composite concrete beams", *The Structural Engineer*, Vol 42, No 6, June 1964.

(18) Teraszkiewicz, J S, "Static and fatigue behaviour of simply supported and continuous composite beams of steel and concrete", PhD Thesis, University of London, 1967.

(19) Ollgaard, J G, Slutter, R G and Fisher, J W, "Shear strength of stud connectors in lightweight and normal-weight concrete", *Engineering Journal AISC*, Vol 8, April 1971, pp 55–64.

(20) Menzies, J B, "CP 117 and shear connectors in steel-concrete composite beams made with normal-density or lightweight concrete", *The Structural Engineer*, Vol 49, No 3, March 1971, pp 137–154.

(21) Osborne-Moss, D M, "Limit states of composite steel-concrete bridges", PhD Thesis, University of London, July 1971.

(22) "Draft Standard specification for the structural use of steelwork in building. Part 4: Design of composite floors with profiled steel sheeting", British Standards Institution, 1978.

(23) *European Recommendations for the design of composite floors with profiled steel sheet*, Published by Constrado, November 1976.

(24) Bresler, B, *Reinforced concrete engineering, Vol 1: Materials, structural elements, safety*, John Wiley & Sons, 1974.

(25) Neville, A M, *Properties of concrete*, Pitman Publishing, Second (Metric) edition 1973, reprinted 1975.

(26) Emerson, M, "Temperature differences in bridges: basis of design requirements", Transport and Road Research Laboratory Report 765, Crowthorne, 1977.

(27) Roik, K, "Methods of prestressing continuous composite girders", Proceedings, Conference on Steel Bridges, June 1968. British Constructional Steelwork Association, London, 1969, pp 75–81.

(28) Stevens, R F, "Encased stanchions and BS 449", *Engineering*, Vol 188, October 1959, p 376.

(29) Stevens, R F, "Encased stanchions", *The Structural Engineer*, Vol 43, February 1965, pp 59–66.

(30) Bondale, D S, "The effect of concrete encasement on eccentrically loaded steel columns", PhD Thesis, University of London, 1962.

(31) Basu, A K and Sommerville, W, "Derivation of formulae for the design of rectangular composite columns", *Proc ICE*, Supplement Volume, 1969, pp 233–280.

(32) Yam, L C P, "Rationalization of structural design—recent developments in EEC Countries", Second Conference on Steel Developments, May 1977, Australian Institute of Steel Construction, Melbourne, 1977.

(33) Virdi, K S and Dowling, P J, "A unified design method for composite columns", CESLIC Report (Civil Engineering Department, Engineering Structures Laboratories, Imperial College), No CC8, July 1975.

(34) Virdi, K S and Dowling, P J, "The ultimate strength of composite columns in biaxial bending", *Proc ICE*, Part 2, Vol 55, Paper 7598, March 1973, pp 251–272.

(35) Wood, R H, "Effective lengths of columns in multi-storey buildings", Building Research Establishment, Current Paper CP 85/74, September 1974.

(36) "Design Manual for Concrete-filled Hollow Section Steel Columns", Cidect Monograph No 1, British Steel Corporation, 1970.

(37) Neogi, P K, Sen, H K and Chapman, J C, "Concrete-filled tubular steel columns under eccentric loading", *The Structural Engineer*, Vol 47, May 1969, pp 187–195.

(38) Sen, H K, "Triaxial effects in concrete-filled tubular steel columns", PhD Thesis, University of London, July 1969.

(39) Yam, L C P, "New developments in theory and method of analysis for inelastic plates and shells", Proceedings of Symposium on "Structural Analysis, Non-linear Behaviour and Techniques", TRRL Supplementary Report 164 UC, 1975.

(40) Noble, P W and Leech, L V, *Design tables for composite steel and concrete beams*, published by Constrado.

(41) Knowles, P R, *Simply supported composite plate girder highway bridge*, published by Constrado, December 1976.

Index